Lecture Notes in Mathematics

Edited by A. Dold and B. Eckmann
Series: Mathematisches Institut der Universität Bonn
Adviser: F. Hirzebruch

601

Modular Functions of one Variable V

Proceedings International Conference,
University of Bonn, Sonderforschungsbereich Theoretische Mathematik
July 2–14, 1976

Edited by
J.-P. Serre and D. B. Zagier

Springer-Verlag
Berlin Heidelberg New York 1977

Editors
Jean-Pierre Serre
Collège de France
75231 Paris Cedex 05
France

Don Bernard Zagier
Mathematisches Institut
der Universität Bonn
Wegelerstr. 10
53 Bonn/BRD

Library of Congress Cataloging in Publication Data
Main entry under title:

Modular functions of one variable V– .

(Lecture notes in mathematics ; 601)
1. Functions, Modular—Congresses. 2. Algebraic number theory—Congresses. I. Serre, Jean-Pierre. II. Zagier, Don Bernard. III. Series: Lecture notes in mathematics (Berlin) ; 601)
QA3.L28 no. 601, etc. [QA343] 510'.8s 77-22148
[515'.9]

AMS Subject Classifications (1970): 10C15, 10D05, 10D25, 12A99, 14H45, 14K22, 14K25

ISBN 3-540-08348-0 Springer-Verlag Berlin Heidelberg New York
ISBN 0-387-08348-0 Springer-Verlag New York Heidelberg Berlin

Printing and binding: Beltz Offsetdruck, Hemsbach/Bergstr.
2141/3140-543210

PREFACE

In view of the increased interest and rapid developments in the theory of modular forms which followed the 1972 conference in Antwerp, it was decided to have a second meeting on the same subject. This meeting was held in Bonn in the summer of 1976 in place of the annual Mathematische Arbeitstagung. The organizers were F. Hirzebruch, J-P. Serre and D. Zagier.

The proceedings of the conference are being published in two parts, as a continuation of the four volumes of proceedings of the Antwerp conference (Lecture Notes Nos. 320, 349, 350 and 476). The present volume is mostly algebraic (congruence properties of modular forms, modular curves and their rational points, etc.), whereas the second volume will be more analytic and also include some papers on modular forms in several variables.

The conference was sponsored by the "Special Research Area in Theoretical Mathematics" (Sonderforschungsbereich 40 der Deutschen Forschungsgemeinschaft) at the University of Bonn, with further support from the Collège de France and the International Mathematical Union. We would like to thank these organizations as well as the National Science Foundation, whose generous financial support made it possible for a large number of mathematicians from the United States to attend the meeting. We would also like to thank W. Kuyk of the University of Antwerp, who placed secretarial help at our disposal for the typing of many of the manuscripts, and the secretaries of the Mathematical Institute of the University of Bonn, especially Mrs. E. Gerber, for the effort they put in to make the conference a success.

Jean-Pierre Serre   Don Zagier

CONTENTS

List of Participants

Y. Amice (Paris)
A.N. Andrianov (Leningrad)
A.O.L. Atkin (Chicago)
P. Barrat (Paris)
B. Beck (Paris)
L. Bégueri (Orsay)
R. Berndt (Hamburg)
P. Berthelot (Rennes)
B. Birch (Oxford)
E. Böhme (Hamburg)
A. Borel (I.A.S.-Princeton)
L. Breen (Rennes)
A. Bremner (Cambridge)
A. Brumer (New York)
J.-L. Brylinski (Paris)
J. Buhler (Harvard)
J.W.S. Cassels (Cambridge)
J. Coates (Cambridge)
H. Cohen (Bordeaux)
H. Cohn (New York)
L. Cohn (Johns Hopkins)
K. Doi (Bonn)
D. Duval (Grenoble)
B. Dwork (Princeton)
P. Ehrlich (Bonn)
M. Eichler (Basel)
J. Elstrodt (Münster)
Y. Flicker (Cambridge)
J.-M. Fontaine (Grenoble)
G. Frey (Saarbrücken)
R.D. Friedman (Harvard)
S. Gelbart (Cornell)
P. Gérardin (Paris)
D. Goldfeld (M.I.T.)
W. Gordon (Philadelphia)
K.B. Gundlach (Marburg)
W. Hammond (Albany,N.Y.)
G. Harder (Bonn)
F.P. Heider (Cologne)
F. Hirzebruch (Bonn)
D. Husemoller (Haverford College)
F.W. Kamber (Illinois)
E. Kani (Heidelberg)
M.L. Karel (Chapel Hill N.C.)
N. Katz (Princeton)
R. Kiehl (Mannheim)
F. Kirchheimer (Freiburg)
H. Klingen (Freiburg)
J. Klingen (Essen)
N. Klingen (Cologne)
P. Kluit (Amsterdam)
N. Koblitz (Harvard)
F.J. Koll (Bonn)
HP. Kraft (Bonn)
D. Kubert (Cornell)
S.S. Kudla (College Park, Md.)
R. Kulle (Göttingen)
W. Kuyk (Antwerp)
H. Lang (Münster)
S. Lang (Yale)
O. Lecacheux (Paris)
W. Li (Harvard)
S. Lichtenbaum (Cornell)
G. Ligozat (Orsay)
J. Masley (Chicago)
C.R. Matthews (Cambridge)
B. Mazur (Harvard)
E. Mendoza (Bonn)
J. Merriman (Kent)
H.B. Meyer (Freiburg)
W.T. Meyer (Bonn)
C.J. Moreno (I.A.S.-Princeton)
H. Naganuma (Bonn)
A. Nobs (Bonn)
J. Oesterlé (Paris)
A. Ogg (Bonn)
M. Ozeki (Okinawa)
S.J. Patterson (Cambridge)
R. Perlis (Bonn)
M. Peters (Münster)
I.I. Piatetskii-Shapiro (College Park)
G. Poitou (Orsay)
C.L. Queen (Cornell)
F.M. Ragab (Cairo)
R.A. Rankin (Glasgow)
M. Raynaud (Tours)
M. Razar (College Park, Md.)
H.L. Resnikoff (U.C. Irvine)
K. Ribet (Princeton)
G. Robert (Paris)
H. Saito (Bonn)
P. Satgé (Caen)
R. Schertz (Cologne)
H. Schiek (Bonn)
T. Schleich (Wuppertal)
W. Schmid (Columbia)
V. Schneider (Mannheim)
J. Schwermer (Bonn)
J.-P. Serre (Paris)
K.-Y. Shih (Ann Arbor, Mich.)
W. Sinnott (Princeton)
J.B. Slater (London)
H. Stark (M.I.T.)
N.M. Stephens (Carditt)
H. Stützer (Cologne)
D. SubRao (Saarbrücken)
H.P.F. Swinnerton-Dyer (Cambridge)
J. Tate (Harvard)
B. Tennison (Cambridge)
A. Terras (San Diego)
C.-H. Tzeng (Taiwan)
L. Van Hamme (Brussels)
G. Van Steen (Antwerp)
J. Vélu (Orsay)
A. Verschoren (Antwerp)
M.-F. Vignéras (Orsay)
G. White (Oxford)
A.J. Wiles (Cambridge)
J. Wolfart (Freiburg)
D. Zagier (Bonn)
H.G. Zimmer (Saarbrücken)

Ramanujan's unpublished work on congruences

by R.A. Rankin

1. Introduction.

The Collected Papers of Srinivasa Ramanujan (1887-1920) [4] contain three papers [10], [11], [12] on congruence properties of partitions. The last of these [12] was published posthumously as a short note and bears the rubric: Extracted from the manuscripts of the author by G.H. Hardy.

In an explanatory footnote Hardy states that the manuscript from which the note is derived is a sequel to Ramanujan's paper [10] and goes on to remark: "The manuscript contains a large number of further results. It is very incomplete, and will require very careful editing before it can be published in full. I have taken from it the three simplest and most striking results, as a short but characteristic example of the work of a man who was beyond question one of the most remarkable mathematicians of his time."

This manuscript, which is unsigned, was, according to J.M. Rushforth [16], sent to Hardy by Ramanujan a few months before the latter's death on 26 April, 1920. It was presumably enclosed with his last letter to Hardy of 12 January, 1920. I shall refer to the manuscript as the MS. It bears the title <u>Properties of</u> $p(n)$ <u>and</u> $\tau(n)$ <u>defined by the equations</u>*

$$\sum_{0}^{\infty} p(n)x^n = \frac{1}{(1-x)(1-x^2)(1-x^3)\cdots},$$

$$\sum_{1}^{\infty} \tau(n)x^n = x\{(1-x)(1-x^2)(1-x^3)\cdots\}^{24}.$$

It is hand-written, divided into 19 sections and occupies 43 foolscap pages, the last nine consisting of additional material to be inserted at indicated earlier points of the text. From remarks in §§16 and 19 it is clear that the MS forms

*not <u>relations</u>, as stated in [16].

the first part of a longer paper. In his obituary notice of Ramanujan [4, p.xxxiv] Hardy rates the MS, together with five other papers (including [9]), as the most remarkable of Ramanujan's contributions to mathematics.

In 1928 the MS was passed on to G.N. Watson, who was then commencing a series of papers devoted to Ramanujan's work; see [13]. After Watson's death in 1965 all his unpublished work on Ramanujan's notebooks, together with the MS and various other documents were, at my suggestion, donated by his widow to Trinity College, Cambridge, of which Hardy, Ramanujan and Watson had been Fellows. A photocopy of the MS is also held by the Ramanujan Institute, Madras.

References to the MS are to be found, not only in [12] and [16], but also in [22] and in Hardy's book on Ramanujan [3, pp.100, 184].

Before Watson's death he presented to the library of the Mathematical Institute of Oxford University his own copy of Ramanujan's second notebook and of other 'various manuscripts' and documents relating to Ramanujan. It is clear that one of these 'various manuscripts', namely Fragment [VII] of [1], is the sequel to the MS. It consists of five sections, numbered §§20-24, and its contents have been described by Birch in his paper; it is concerned solely with the partition function. I am indebted to Dr Rushforth, who sent me a photocopy of his own hand-written copy of this Fragment (made when he was a research student of Watson), for providing the information that has enabled me to identify Birch's Fragment [VII] with the second part of the MS. The copy of the second notebook from which Watson made his transcript is in the library of Trinity College, Cambridge. The original is presumably in the Ramanujan Institute of the University of Madras.

The main purpose of the present paper is to describe Ramanujan's work in the MS on congruences satisfied by his function $\tau(n)$. However, it may be of interest to state that it contains a few results concerning the partition function $p(n)$ additional to those published in [12].

For example, proofs of the congruences modulis 25, 49 and 121 stated in [11] are given; the method of proof differs from that sketched in [10] and is, doubtless, the new method referred to in [11]; see [16, §§7, 8] and also [1]. The MS also contains numerical information on the number of values of $n \leq 200$ for which $p(n)$ is congruent to various residues modulis 2, 3, 5, 7 and 11.

Moreover, the partition function is considered modulo 13; it is stated that, for any positive integers λ and n with $\lambda + n$ even, the residue of

$$p\left(\frac{13^{\lambda}(2n - 1) + 1}{24}\right)$$

modulo 13 can be completely ascertained. For example, the congruence

$$p\left(\frac{11.13^{\lambda} + 1}{24}\right) + 2^{(5\lambda-3)/2} \equiv 0 \pmod{13}$$

is one of fourteen similar results included. Results of this type have since been obtained by M. Newman [8] for λ and n both odd. Other results on partitions from the MS will be found in [16, §9].

2. Congruences of the first kind.

I use Watson's division of congruences into those of 'erster und zweiter Art Those of the first kind are congruences of the kind considered in Swinnerton-Dyer's Antwerp lecture notes [21]. The methods developed by Serre [18], Deligne [2] and Swinnerton-Dyer use ℓ-adic representation theory to show that, for $\tau(n)$, there are exactly six prime moduli of two different types, namely (i) $\ell = 2, 3, 5, 7, 691$ and (ii) $\ell = 23$, for which congruences can hold. For the first five of these 'exceptional primes' the congruences can be expressed in the form

$$\tau(n) \equiv n^{m} \sigma_{11-2m}(n) \pmod{\ell}, \tag{1}$$

for certain fixed values of the integer m and for all positive integers prime to ℓ. For $\ell = 23$ the congruence is

$$\tau(n) \equiv 0 \pmod{\ell}, \tag{2}$$

whenever n is a quadratic non-residue modulo ℓ. Here, as usual,

$$\sigma_\nu(n) = \sum_{d|n, d>0} d^\nu$$

for real ν.

Various refinements of (1) are possible, with ℓ replaced by a power of ℓ and different divisor functions on the right-hand side. Moreover, for each of the five other cusp-form coefficients $\tau_r(n)$, where $r = 2, 3, 4, 5, 7$, there is a similar division of exceptional primes into two (or possibly three, when $r = 2$) types. Here

$$\sum_{n=1}^{\infty} \tau_r(n) q^n = \Delta(z) E_{2r}(z), \tag{3}$$

where, as usual, Δ is the discriminant function, E_{2r} is the Eisenstein series of weight $2r$ and constant term 1 and $q = e^{2\pi i z}$; we may also take $r = 0$, in which case $E_0(z) = 1$ and $\tau_0(n) = \tau(n)$.

Ramanujan was aware that the coefficients $\tau_r(n)$ $(r = 2, 3, 4, 5, 7)$ have similar multiplicative properties to $\tau(n)$ and this is mentioned in the MS, where the corresponding Euler products for the associated Dirichlet series are set down.

The following congruences for $\tau(n)$ are given in the MS. Since Ramanujan's death these results have been proved or improved by other mathematicians, the most recent results being listed in [21, formulae (2)-(7)]. The reference numbers attached to the congruences in the MS are quoted on the left.

(12.1) $\tau(n) \equiv n^3\sigma_1(n) \pmod{16}$, $\tau(n) \equiv n^3\sigma_5(n) \pmod{32}$, (4)

(12.3) $\tau(n) \equiv n^2\sigma_1(n) \pmod{9}$, $\tau(n) \equiv n^2\sigma_7(n) \pmod{27}$, (5)

(2.1) $\tau(n) \equiv n\sigma_1(n) \pmod{5}$, (6)

(4.2) $\tau(n) \equiv n\sigma_9(n) \pmod{25}$, (7)

(6.2) $\tau(n) \equiv n\sigma_3(n) \pmod{7}$, (8)

(12.7) $\tau(n) \equiv \sigma_{11}(n) \pmod{691}$. (9)

These hold for all positive integers n. Proofs are given although in many cases the details are suppressed or covered by statements such as "it is easy to see that ...". Moreover, at the end of §12 (p.8 of insertions) it is shown that

$$\tau(n) \equiv \sigma_{11}(n) \pmod{256} \quad (n \text{ odd}). \tag{10}$$

After his proof of (7) Ramanujan remarks: "It appears that, if k be any positive integer, it is possible to find two integers a and b such that

(4.3) $$\tau(n) - n^a\sigma_b(n) \equiv 0 \pmod{5^k} \tag{11}$$

if n is not a multiple of 5. Thus, for instance,

(4.4) $$\tau(n) - n^{41}\sigma_{29}(n) \equiv 0 \pmod{125}, \tag{12}$$

if n is not a multiple of 5. I have not yet proved these results."

The congruence (12) is true and, since $d^{100} \equiv 1 \pmod{125}$ when $(d, 5) = 1$, is equivalent to Serre's congruence [18], [21]

$$\tau(n) \equiv n^{-30}\sigma_{71}(n) \pmod{5^3} \quad (n, 5) = 1.$$

However, Ramanujan's conjecture (11) is false for $k \geq 4$. For 443 is prime and its powers are congruent to ± 1, $\pm 443 \pmod{5^4}$. But, from [23],

$$\tau(443) \equiv -58 \pmod{5^4}.$$

Hence no integers a and b exist for which (11) holds with $k \geq 4$ and $n = 443$.

Ramanujan also considers congruences modulo 49, but his final result (formula (8.6) of the MS) involves another cusp-form coefficient as well as divisor functions.

The first published proof of (9) was given by Wilton [24], who draws attention to the fact that it follows immediately from equation (1.63) of [12] and that the same formula occurs as formula 6 of Table I of [9]. His statement that Ramanujan does not seem to have noticed this fact is not borne out by the MS where the congruence is derived in precisely this way.

The congruence (2) (with $\ell = 23$) was stated by Ramanujan in [11]. In the MS the cases when n is not a quadratic non-residue are also considered. In 1928 Wilton [25] determined completely the residues of $\tau(n)$ modulo 23. Both Wilton and Ramanujan reduce the problem to the consideration of the coefficients of $\eta(z)\eta(23z)$, where η is Dedekind's function, but Ramanujan bases his argument (in §17 of the MS) on the fact that the associated Dirichlet series of this cusp-form of weight 1 has an Euler product over primes p of three different types: (a) p a non-residue modulo 23, (b) p a quadratic residue modulo 23, but not of the form $23a^2 + b^2$, and (c) the remaining primes. For this fact he offers no proof.

In fact these properties can be deduced from the fact that $\eta(z)\eta(23z)$ is a newform for the group $\Gamma_o(23)$ with multiplier system given by the character

$$\chi(n) = \left(\frac{n}{23}\right)$$

(Legendre symbol), together with Wilton's determination of its Fourier coefficients c_p for prime p.

The prime 11, although not exceptional for $\tau(n)$, is so for $\tau_r(n)$, when $r = 2, 3, 4$ and 7. In the MS Ramanujan remarks that it is easy to show that

(18.3) $$\tau_r(n) \equiv n\sigma_{2r-1}(n) \pmod{11} \quad (r = 2, 3, 4). \tag{13}$$

See the table on p.32 of [21], which includes these among other congruences.

In Rushforth's (unpublished) Ph.D. thesis [15] can be found proofs of congruences equivalent to all those listed in Swinnerton-Dyer's table of congruences satisfied by $\tau_r(n)$ $(r > 0)$; for $\ell = 2$ and 3 the moduli are, in several cases, powers of ℓ. Dr Rushforth has also drawn my attention to the fact that one of the congruences in the table, namely

$$\tau_2(n) \equiv \sigma_{15}(n) \pmod{3617},$$

was given in a disguised form by Wilton [24, formula (5.3)]; see also, for this modulus, Serre [18, §5.3].

For references to the many other papers on congruences satisfied by $\tau(n)$ see LeVeque [7].

3. Congruences of the second kind.

These are congruences that state that a given arithmetical function $a(n)$ is divisible by a fixed positive integer ℓ for almost all n; i.e.

$$A_\ell(x) = \sum_{\substack{n\le x\\ \ell\nmid a(n)}} 1 = o(x) \tag{14}$$

as $x \to \infty$. It is possible in some cases to replace (14) by more precise estimates, such as

$$A_\ell(x) \sim \frac{A_\ell x}{(\log x)^{\alpha_\ell}}, \tag{15}$$

where A_ℓ and α_ℓ are positive constants.

The MS contains various results of this type for $\tau(n)$ and different values of ℓ. To attack the problem Ramanujan defines t_n for $n \ge 1$ by

$$t_n = 0 \quad \text{if} \quad \ell|\tau(n), \quad t_n = 1 \text{ otherwise}, \tag{16}$$

and shows that, when ℓ is prime, the Dirichlet series $\sum_{n=1}^{\infty} t_n n^{-s}$ possesses an Euler product. He then has to consider

$$A_\ell(x) = \sum_{n\le x} t_n .$$

When $\ell = 3, 5, 7$ and 23 he evaluates the Euler products explicitly. He then states (except for $\ell = 3$) that it can be shown by elementary methods that (14) holds, and by transcendental methods that (15) holds. For $\ell = 5$ (the first case considered) it is stated that the proof of (14) is "quite elementary and very easy similar to that for showing that $\pi(x) = o(x)$". An incomplete reference to Landau's Primzahlen [6] is given as a footnote (presumably pp.69-71).

For $\ell = 3$, 7 and 23 the values of the constant A_ℓ in (15) are given; that his values of A_3 and A_7 are correct I have confirmed from [14]. He also gives the correct values for α_ℓ, namely $\frac{1}{2}$, $\frac{1}{4}$, $\frac{1}{2}$ and $\frac{1}{2}$ for $\ell = 3, 5, 7$ and 23, respectively. The case $\ell = 691$ is also considered although A_{691} is not evaluated explicitly; the correct value $\alpha_{691} = 1/690$ is given.

In a final most interesting section (§19) Ramanujan reverts to the problem in a more general form, and I quote in detail from the MS:

"Suppose that $\varpi_1, \varpi_2, \varpi_3, \ldots$ are an infinity of primes such that

(19.1) $$\frac{1}{\varpi_1} + \frac{1}{\varpi_2} + \frac{1}{\varpi_3} + \cdots \tag{17}$$

is a divergent series and also suppose that $a_2, a_3, a_5, a_7, \ldots$ assume some or all of the positive integers (including zero) but that $a_{\varpi_1}, a_{\varpi_2}, a_{\varpi_3}, \ldots$ never assume the value unity. Then it is easy to show that the number of numbers of the form

(19.2) $$2^{a_2}.3^{a_3}.5^{a_5}.7^{a_7}. \ldots$$

not exceeding n is of the form

(19.3) $$o(n).$$

In particular if a_ϖ never assumes the value unity for all prime values of ϖ of the form

(19.4) $$\varpi \equiv c \pmod{k}$$

where c and k are any two integers which are prime to each other then the number of numbers of the form (19.2) is of the form

(19.5) $$o(n)$$

and more accurately is of the form

(19.6) $$O\,\frac{n}{(\log n)^{1/(k-1)}}$$

where k is the same as in (19.4).

Thus for example if s be an odd positive integer the number of values of n not exceeding n for which $\sigma_s(n)$ is <u>not</u> divisible by k where k is any positive integer is of the form

$$o(n) \tag{19.7}$$

and more accurately is of the form

$$O\,\frac{n}{(\log n)^{1/(k-1)}}.$$

For if n be written in the form

$$2^{a_2}.3^{a_3}.5^{a_5}.7^{a_7}.\ \ldots$$

then we have

$$\sigma_s(n) = \prod_p \frac{p^{s(1+a_p)} - 1}{p^s - 1}, \quad p = 2, 3, 5, 7, 11, \ldots .$$

Since s is odd $\sigma_s(n)$ is divisible by k at any rate when $a_p = 1$ for all values of p of the form

$$p \equiv -1 \pmod k)$$

and hence the results stated follow. Thus we see that, if s is odd $\sigma_s(n)$ is divisible by any given integer for almost all values of n.

It follows from all these and the formulae in §§ † that

$$\tau(n) \equiv 0 \ (\mathrm{mod}\ 2^5.3^3.5^2.7^2.23.691) \tag{19.9}$$

for almost all values of n.

It appears that $\tau(n)$ is almost always divisible by any power of 2, 3 and 5. It also appears from §10 * that there are reasons to suppose that $\tau(n)$ is almost always divisible by 11 also. But I have no evidence at present to say anything about the other powers of 7 and other primes one way or the other.

Among the values of $\tau(n)$ multiples of 2, 3, 5, 7 and 23 are very numerous

---

† The missing references appear to be to §§4, 6, 8, 12, 17.

* Reference supplied by me. R.A.R.

from the beginning but multiples of 691 begin at a very late stage. For instance $\tau(n)$ is divisible by 23 for 132 values of n not exceeding 200 while the first value of n for which $\tau(n)$ is divisible by 691 is 1381 and this is the only† value of n among the first 5000 values."

This extract displays Ramanujan's remarkable insight. In the case of exceptional primes of the first type it is clear from (1) that the problem is essentially one of the divisibility of divisor functions. It is interesting that Watson, who had the MS before him when writing his paper [22], obtained in his Hauptsatz 2 only an O-result and not an asymptotic result of the form (15) for the divisor function $\sigma_s(n)$ modulo k; he could therefore only obtain an O-result for $\tau(n)$ modulo 691.

It is also curious that, in §2 of [22], Watson states as a 'Vermutung von Ramanujan' a weaker form of the result that Ramanujan, in the extract above, states that it is easy to show.

The results concerning the divisibility of divisor functions have all, of course, since been proved and extended (in ignorance of the contents of the MS) by Rankin [14], Scourfield [17] and others.

However, this work has now been overtaken by the new results of Serre [19], [20], which are not restricted to divisor functions or to exceptional primes and in which general conditions like (17) play an important role. Serre's work shows that Ramanujan was correct in his conjecture regarding the prime 11. In support of this conjecture Ramanujan states, in §10 of the MS (page 4 of the material to be inserted), that there are primes p, such as 19, 29, ... for which $\tau(pn) \equiv 0 \pmod{11}$ and that it is very likely that the sum of the reciprocals of these primes is divergent.

It is of interest, in conclusion, to consider what Ramanujan had in mind when he stated that results of the form (14) can be obtained by transcendental methods. Although his knowledge of complex function theory may have been slight,

---

† This is true only for prime n.

I believe that he knew enough to realize that Landau's contour-integration method [5], which is basic to the analysis carried out in [14], [17], [19], [20], [22], applied to his problem. It would be surprising if he had not known of Landau's work. For, in one of his first letters to Hardy [4, p.xxiv], [3, p.8] in 1913 before he came to England, he gave the approximation

$$K\int_A^x \frac{dt}{\sqrt{(\log t)}}$$

for the number of integers between A and x that are sums of two non-negative squares. That this result had been proved by Landau [5] must have been drawn to his attention by Hardy, who remarks also that the integral has no advantage as an approximation over the simpler function

$$\frac{Kx}{\sqrt{(\log x)}}$$

and that Ramanujan had been deceived by a false analogy with the problem of the distribution of primes. It is of interest that in the MS he does the same and gives integrals as the main terms in his approximations (formulae (2.8), (6.7), (17.8)).

REFERENCES

1. B.J. Birch, A look back at Ramanujan's notebooks, Math. Proc. Cambridge Philos. Soc. 78(1975), 73-79.

2. P. Deligne, Formes modulaires et représentations ℓ-adiques. Séminaire Bourbaki, 355, February 1969.

3. G.H. Hardy, Ramanujan; twelve lectures on subjects suggested by his life and work (Cambridge, 1940).

4. G.H. Hardy, P.V. Seshu Aiyar and B.M. Wilson (editors), Collected papers of Srinivasa Ramanujan (Cambridge, 1927).

5. E. Landau, Über die Einteilung der positiven ganzen Zahlen in vier Klassen nach der Mindestzahl der zu ihrer additiven Zusammensetzung erforderlichen Quadrate, Arch. Math. Phys. (3) 13(1908), 305-312.

6. E. Landau, Handbuch der Lehre von der Verteilung der Primzahlen (Leipzig, 1909).

7. W.J. LeVeque, Reviews in number theory, Amer. Math. Soc., Providence, R.I., 1974, Vol.2, F35.

8. M. Newman, Congruences for the coefficients of modular forms and some new congruences for the partition function, Canad. J. Math. 9(1957), 549-552.

9. S. Ramanujan, On certain arithmetical functions, Trans. Cambridge Philos. Soc. 22(1916), 159-184; no 18 in [4].

10. S. Ramanujan, Some properties of p(n), the number of partitions of n, Proc. Cambridge Philos. Soc. 19(1919), 207-210; no 25 in [4].

11. S. Ramanujan, Congruence properties of partitions, Proc. London Math. Soc.(2) 18(1920), xix-xx; no 28 in [4].

12. S. Ramanujan, Congruence properties of partitions, Math. Z. 9(1921), 147-153; no 30 in [4].

13. R.A. Rankin, George Neville Watson (obituary notice), J. London Math. Soc. 41(1966), 551-565.

14. R.A. Rankin, The divisibility of divisor functions, Proc. Glasgow Math. Assoc. 5(1961), 35-40.

15. J.M. Rushforth, Congruence properties of the partition function and associated functions, Ph.D. thesis, University of Birmingham, 1950.

16. J.M. Rushforth, Congruence properties of the partition function and associated functions, Proc. Cambridge Philos. Soc. 48(1952), 402-413.

17. E.J. Scourfield, On the divisibility of $\sigma_\nu(n)$, Acta Arith. 10(1964), 245-285.

18. J.-P. Serre, Une interprétation des congruences relatives à la fonction τ de Ramanujan. Séminaire Delange-Pisot-Poitou: 1967/68, Théorie des Nombres, Fasc. 1, Exp.14, 17 pp.

19. J.-P.Serre, Divisibilité des coefficients des formes modulaires de poids entier, C.R. Acad. Sci. Paris Sér.A 279(1974), 679-682.

20. J.-P. Serre, Divisibilité de certaines fonctions arithmétiques. Séminaire Delange-Pisot-Poitou: 1974/75, Théorie des Nombres, Exp.20, 28pp.

21. H.P.F. Swinnerton-Dyer, On ℓ-adic representations and congruences for coefficients of modular forms, Modular functions of one variable, III, Lecture Notes in Mathematics, 350(1973), 1-55.

22. G.N. Watson, Über Ramanujansche Kongruenzeigenschaften der Zerfallungsanzahlen I, Math.Z. 39(1935), 712-731.

23. G.N. Watson, A table of Ramanujan's function $\tau(n)$, Proc. London Math. Soc.(2) 51(1948), 1-13.

24. J.R. Wilton, On Ramanujan's arithmetical function $\Sigma_{r,s}(n)$, Proc. Cambridge Philos. Soc. 25(1929), 255-264.

25. J.R. Wilton, Congruence properties of Ramanujan's function $\tau(n)$, Proc. London Math. Soc.(2) 31(1929), 1-10.

University of Glasgow,
Glasgow.

GALOIS REPRESENTATIONS ATTACHED TO EIGENFORMS WITH NEBENTYPUS

by Kenneth A. Ribet

GALOIS REPRESENTATIONS ATTACHED TO EIGENFORMS WITH NEBENTYPUS

Kenneth A. Ribet*

This paper is the written version of a set of survey lectures on the subject of ℓ-adic representations attached to modular forms. My aim at the conference was to summarize what one knows about these representations and what one can prove about modular forms by using them. In writing up the lectures, I gradually supressed subjects which were already treated in print (e.g., representations attached to forms of weight 1, "mod ℓ" representations) and began drifting toward things that are new, or at least not reported on elsewhere.

The main theme of this paper can most easily be described in the context of results of Serre/Swinnerton-Dyer [20,28] and the author [14] on the ℓ-adic representations attached to eigenforms of level 1, i.e., on $SL_2\mathbb{Z}$. There the idea was to determine the images of these representations as exactly as possible, the key point being that a "small" image means a congruence "mod ℓ" involving the coefficients of modular forms. Here we work instead with eigenforms of arbitrary level $N \geq 1$ and try to describe the images "up to finite groups," as measured by the ℓ-adic Lie algebras of the representations. What then becomes important is the possibility of equalities, as opposed to congruences, involving coefficients. More specifically, it is crucial to study relationships between a given form and its "twists" by Dirichlet characters.

Actually, the Lie algebras are not explicitly mentioned until the final §, and the initial §'s are all devoted to simple questions involving the eigenvalues for a given eigenform and the resulting representations. The first § is a review of some facts about eigenforms with character, and especially new forms. In §2 we state the theorems associating ℓ-adic representations to such eigenforms and begin to

*Research partially supported by Sloan Foundation grant BR-1689.

establish some elementary properties of these representations. In particular, we prove that the representations attached to a cusp form are irreducible, and this implies that they are unique up to isomorphism.

The next two §'s answer the question: When are the ℓ-adic representations associated to a cusp form abelian on some open subgroup of the Galois group of $\mathbb{Q}$? Aside from the case of weight 1 when the answer is "always," it turns out that the form must be associated to a Grössencharacter of an imaginary quadratic field; we then say that the form has complex multiplication by that field.

In the final § we take up the question of determining the ℓ-adic Lie algebras $\mathfrak{g}_\ell$ for a cusp form in weight $k > 1$ which does not have complex multiplication. We are not able to do this if the form is (in some non-trivial way) a "twist" of one of its conjugates, in which case we say the form has "extra twists." If the form does not have extra twists, however, we exhibit a Lie algebra $\mathfrak{g}$ over $\mathbb{Q}$ such that $\mathfrak{g}_\ell = \mathfrak{g} \otimes \mathbb{Q}_\ell$ for all primes ℓ. When the weight of the form is even, $\mathfrak{g}$ is essentially a matrix algebra, but if the weight is odd, we get a definite quaternion algebra instead. (The quaternionic structure is due essentially to the fact that the classical operator $W = \begin{pmatrix} 0 & -1 \\ N & 0 \end{pmatrix}$ "commutes" with Hecke operators according to the formula

$$T_p W = W T_p^t,$$

where ^{t} represents transpose with respect to the Petersson product.) It is presumably possible to give a determination of the $\mathfrak{g}_\ell$ for forms with extra twists by constructing operators as in [27, Prop. 9]. However, this determination has not yet been carried out.

§1. Review of Eigenforms with Nebentypus.

Let $N, k \geq 1$ be integers, and let

$$S = S_k(\Gamma_1(N))$$

be the complex vector space of cusp forms of weight k on $\Gamma_1(N)$. For $\begin{pmatrix} a & b \\ c & d \end{pmatrix} \in \Gamma_0(N)$ the map

$$f \mapsto f|\begin{pmatrix} a & b \\ c & d \end{pmatrix}$$

defines an endomorphism R_d of S which depends only on $d \bmod N$. We thus get an action of $(\mathbb{Z}/N\mathbb{Z})^*$ on S. Since $(\mathbb{Z}/N\mathbb{Z})^*$ is a finite group, there is a decomposition

$$S = \bigoplus_{\varepsilon} S(\varepsilon),$$

where ε runs over the set of characters $(\mathbb{Z}/N\mathbb{Z})^* \to \mathbb{C}^*$ and where $S(\varepsilon)$ is the set of forms $f \in S$ which satisfy

$$f|R_d = \varepsilon(d)f$$

for all $d \in (\mathbb{Z}/N\mathbb{Z})^*$. These are the forms of Nebentypus ε. As is well known, $S(\varepsilon) = 0$ unless $\varepsilon(-1) = (-1)^k$, i.e., unless ε and k have the same parity.

For $n \geq 1$ prime to N, we have the operation on S of the nth Hecke operator T_n. The T_n commute with each other and with the R_d; in particular, they preserve the spaces $S(\varepsilon)$. If $p \nmid N$ is a prime, and if

$$f = \sum_{n=1}^{\infty} a_n q^n \in S(\varepsilon),$$

then

$$f|T_p = \sum_{n=1}^{\infty} a_{pn} q^n + p^{k-1} \varepsilon(p) \sum_{n=1}^{\infty} a_n q^{pn}.$$

The T_n for n prime to N may be expressed in terms of the T_p and R_d by well-known formulas, so that the algebra generated in $\mathrm{End}(S)$ by the T_p and R_d contains all the T_n with n prime to N.

The space S carries the Petersson product, a non-degenerate Hermitian product on S. If ^{t} denotes "transpose" with respect to this product, then

$$(R_d)^t = R_{d^{-1}} \quad \text{(the inverse computed } \bmod N\text{)},$$

and

$$(T_p)^t = R_{p^{-1}} T_p.$$

We see from these formulas that the R_d and T_p commute with their transposes; hence they are all diagonalizable on S. Since they commute with each other, they may be simultaneously diagonalized.

In other words, each $S(\varepsilon)$ has a basis consisting of eigenvectors for the Hecke operators T_p, usually called eigenforms. Each eigenform has a "package" of eigenvalues attached to it; these are summarized by the Nebentypus character ε such that $f \in S(\varepsilon)$ and by the complex numbers c_p $(p \nmid N)$ such that

$$f | T_p = c_p \cdot f.$$

From the formula giving T_p^t in terms of $R_{p^{-1}}$ and T_p we derive the relation

$$c_p = \bar{c}_p \varepsilon(p)$$

among the eigenvalues. (The bar denotes complex conjugation.)

Let $K = K_f$ be the subfield of $\mathbb{C}$ generated by all the numbers $\varepsilon(d)$ and c_p. It is a fundamental fact that K is a number field, i.e., a finite extension of $\mathbb{Q}$. This comes from the existence of a $\mathbb{Q}$-subspace S_o of S which is stable under the T_p and R_d and which generates S over $\mathbb{C}$. (We can take S_o to be the set of all forms $f \in S$ whose q-expansion coefficients are rational.) The existence of a $\mathbb{Q}$-form of S shows further that the eigenvalue sets $(\{c_p\}, \varepsilon)$ are permuted by automorphisms of $\mathbb{C}$. More precisely, if σ is such an automorphism and if $f = \Sigma a_n q^n$ is an eigenform with Nebentypus ε and eigenvalues c_p, then

$$f^\sigma = \Sigma \sigma(a_n) q^n$$

lies in $S(\sigma(\varepsilon))$ and satisfies

$$f^\sigma | T_p = \sigma(c_p) f^\sigma$$

for all $p \nmid N$.

Newforms (see [1], [2], [11], [12]).

Clearly the eigenvalue set attached to any eigenform $f \in S$ is also attached to any complex multiple of f. We call the system $(\{c_p\}, \varepsilon)$ primitive if all eigenforms $f \in S$ with this system are scalar multiples of one another.

Theorem (1.1). Let $(\{c_p\}, \varepsilon)$ be a primitive system of eigenvalues. Then there exists a (unique) eigenform

$$f = \sum_{n=1}^{\infty} a_n q^n \in S$$

with this system of eigenvalues which satisfies $a_1 = 1$. For this form we have $a_p = c_p$ for all $p \nmid N$.

One says that f is a newform on $\Gamma_1(N)$ of weight k and Nebentypus ε. The newforms are special (because $a_p = c_p$) eigenforms which in fact give all eigenvalue systems:

Theorem (1.2). Let $(\{c_p\}, \varepsilon)$ be an eigenvalue system on S. Then there exists a divisor M of N such that $(\{c_p\}, \varepsilon)$ is attached to a newform of weight k on $\Gamma_1(M)$.

Implicit in the above statement is the fact that a form on $\Gamma_1(M)$ is a fortiori a form on $\Gamma_1(N)$ when $M|N$: the "identity" map on forms furnishes an injection

$$S_k(\Gamma_1(M)) \hookrightarrow S_k(\Gamma_1(N))$$

which commutes with the operators R_d for $(d, N) = 1$ and T_p for $p \nmid N$. Actually, there are several such maps, one for each divisor of N/M. Namely, if t divides N/M, then the map

$$i_{t,M} : \Sigma a_n q^n \mapsto \Sigma a_n q^{tn}$$

is such an embedding.

The theory of newforms shows how all eigenforms on S may be built out of the newforms on $\Gamma_1(M)$ as M runs over the divisors of N. Let $M|N$, and let

$$f_{1M}, \ldots, f_{r_M M}$$

be the newforms of weight k on $\Gamma_1(M)$. For each j, $1 \leq j \leq r_M$, the forms

$$i_{t,M}(f_{jM}) \text{ for } t \text{ dividing } M/N$$

are eigenforms on S with a common eigenvalue system: that derived from f_{jM}. We let S_{jM} be the subspace of S spanned by these forms.

Theorem (1.3). We have

$$S = \bigoplus_{\substack{M|N \\ 1 \leq j \leq r_M}} S_{jM}.$$

The S_{jM} are precisely the distinct eigenspaces attached to the various eigenvalue systems on S.

We conclude this section by mentioning other properties of eigenforms that we will use.

Dirichlet Series.

Let $f = \sum_{n=1}^{\infty} a_n q^n$ be a newform of weight k and Nebentypus ε on $\Gamma_1(N)$. Let

$$L_f(s) = \sum_{n=1}^{\infty} a_n n^{-s}$$

be the associated Dirichlet series. Then

$$L_f(s) = \prod_p (1 - a_p p^{-s} + \varepsilon(p) p^{k-1-2s})^{-1},$$

with the product running over all primes p. (If $p|N$, we put $\varepsilon(p) = 0$.) The series L_f converges for $\operatorname{Re} s >> 0$ and may be analytically continued to a holomorphic function of s. Set

$$\Lambda_f(s) = N^{s/2}(2\pi)^{-s}\Gamma(s)L_f(s),$$

and similarly define $\Lambda_{\bar{f}}(s)$ for the "complex conjugate" form

$$\bar{f} = \sum_{n=1}^{\infty} \bar{a}_n q^n .$$

Then we have the functional equation

$$\Lambda_f(s) = A \cdot \Lambda_{\bar{f}}(k-s)$$

for some complex number A. (This follows from [24, Th. 3.66] because $\bar{f}$ is a multiple of the form denoted $f|[\tau]_k$ in that theorem.)

Petersson Conjecture.

If $f = \Sigma a_n q^n$ is a cusp form of weight k on $\Gamma_1(N)$, then Rankin [13, Th. 2, p. 358] proved that $|a_n| = O(n^{k/2-1/5})$ as $n \to \infty$. (He proved this more generally for the Fourier coefficients of any cusp form on a congruence subgroup of $SL_2(\mathbb{Z})$, and Selberg [17] extended the method to cusp forms on an arbitrary subgroup of finite index in $SL_2(\mathbb{Z})$.) By applying this estimate with f a newform, we see from (1.1) that the eigenvalues c_p of any eigenform in S satisfy $|c_p| = O(p^{k/2-1/5})$. Recently Deligne has shown [3], [4] for $k \geq 2$ that we have $|c_p| \leq 2p^{(k-1)/2}$ (the Petersson conjecture), and Deligne and Serre [5] established the same estimate for $k = 1$.

§2. The ℓ-adic Representations.

We first fix some terminology concerning the Galois group of $\mathbb{Q}$. Choose an algebraic closure $\bar{\mathbb{Q}}$ of $\mathbb{Q}$, and let $G = G_{\mathbb{Q}}$ be the Galois group $Gal(\bar{\mathbb{Q}}/\mathbb{Q})$. Let p be a prime. Then the choice of a place of $\bar{\mathbb{Q}}$ lying over p determines for p a decomposition group $D \subset G$ and its inertia subgroup $I \subset D$. The quotient D/I is canonically isomorphic to the Galois group $Gal(\bar{\mathbb{F}}_p/\mathbb{F}_p)$ of residue fields, topologically generated by the (Frobenius) automorphism

$$x \mapsto x^p$$

of $\bar{\mathbb{F}}_p$. A Frobenius element in D is any element $F_p \in D$ which maps to this generator in $\mathrm{Gal}(\overline{\mathbb{F}}_p/\mathbb{F}_p)$. A Frobenius element for p in G is any element of G which is a Frobenius element for some decomposition group D for p. Since all decomposition groups for p are conjugate in G, the Frobenius elements for p are the conjugates of the Frobenius elements in a fixed D. We recall that a homomorphism

$$\rho : G \to A \quad (A \text{ any group})$$

is said to be unramified at p if ρ vanishes on the inertia subgroup of one (and hence each) decomposition group for p. If ρ is unramified at p, all Frobenius elements for p in G map to a single conjugacy class in A.

Now let $N, k \geq 1$ be as in §1, and suppose that $f \in S$ is an eigenform as in §1. Let $(\{c_p\}, \varepsilon)$ be the associated system of eigenvalues. Let K_f be the subfield of $\mathbb{C}$ generated by the c_p and the values of ε; as mentioned above, K_f is a number field. Let K be any subfield of $\mathbb{C}$ which contains K_f and which is finite over $\mathbb{Q}$.

Let ℓ be a prime.

Theorem (2.1). There exists a (continuous) representation

$$\rho_\ell : G \to GL(2, K \otimes \mathbb{Q}_\ell)$$

with the following property: If $p \nmid \ell N$ is a prime, then ρ_ℓ is unramified at p, and the image under ρ_ℓ of any Frobenius element for p is a matrix with trace c_p and determinant $\varepsilon(p)p^{k-1}$.

This result was in fact proved as three different theorems. In case $k = 2$, the existence of ρ_ℓ is a "classical" fact in that ρ_ℓ arises from the action of G on ℓ-power division points of an abelian variety over $\mathbb{Q}$. One attaches to f a factor J_f of the Jacobian variety $J_1(N)$ of the modular curve associated to $\Gamma_1(N)$, and the representation space of ρ_ℓ is then the $\mathbb{Q}_\ell$-adic Tate module attached to J_f. The link between this Tate module and the form f is provided by the Eichler-Shimura relation expressing the endomorphism of J_f arising from the

Hecke operator T_p in terms of the Frobenius endomorphism of the reduction at p of J_f. For details, see [24, Ch. 7] and [27].

In the almost-general case $k \geq 2$, the existence of ρ_ℓ was conjectured by Serre (see [18]) and proved by Deligne [3], who showed that the (dual of the) representation occurs in the ℓ-adic analogue of the "Eichler cohomology" group constructed for forms of weight k on $\Gamma_1(N)$.

The remaining case $k = 1$ was treated by Deligne and Serre [5], whose construction depended on the previous results of Deligne in the case $k \geq 2$. More recently Koike [9], following an idea of Shimura, has shown that the Deligne-Serre arguments may be used to produce the weight 1 representations from weight 2 representations; hence the representations in weight 1 may be constructed without using ℓ-adic cohomology.

A final remark about the actual construction of the ρ_ℓ is that we essentially won't use it in what follows. Our point of view is the following:
We are given a 2-dimensional space V_ℓ over $K \otimes \mathbb{Q}_\ell$ on which G acts according to certain axioms, and we derive facts about f (and about ρ_ℓ) from these axioms. However, we do use in the last § a certain additional fact that follows from the construction: V_ℓ carries the action of an operator $W = W_N$ which is analogous to the endomorphism of S given by the operator $\begin{pmatrix} 0 & -1 \\ N & 0 \end{pmatrix}$. Since we do not need W until the end, we postpone our discussion of it.

One says that the representation ρ_ℓ is the ℓ-adic representation attached to f. Its uniqueness (up to isomorphism) will be proved later.

Variant: The λ-adic representations.

Since $K \otimes \mathbb{Q}_\ell$ is the product $\prod_\lambda K_\lambda$ of the various completions of K at the primes λ of K lying over ℓ, we have a decomposition

$$GL(2, K \otimes \mathbb{Q}_\ell) = \prod_{\lambda | \ell} GL(2, K_\lambda).$$

For each λ the composition

$$\rho_\lambda : G \xrightarrow{\rho_\ell} GL(2, K\otimes\mathbb{Q}_\ell) \to GL(2, K_\lambda)$$

is called the λ-adic representation attached to f. The usefulness of ρ_λ arises from the fact that ρ_λ is a representation of G over a field. Since

$$\rho_\ell = \bigoplus_{\lambda|\ell} \rho_\lambda,$$

each ℓ-adic representation may be recovered from its λ-adic components.

Determinants. For each $d \geq 1$, let

$$\varphi_d : G \to (\mathbb{Z}/d\mathbb{Z})^*$$

be the character giving the action of G on the group of dth roots of unity in $\overline{\mathbb{Q}}$. As is well known, φ_d is unramified at each $p \nmid d$ and maps a Frobenius F_p for p to the image of p in $(\mathbb{Z}/d\mathbb{Z})^*$. If ℓ is a prime, the limit

$$\chi_\ell = \varprojlim_n \varphi_{\ell^n}$$

is a (continuous) character $G \to \mathbb{Z}_\ell^*$, called the ℓ-adic cyclotomic character. Since $K\otimes\mathbb{Q}_\ell$ is a $\mathbb{Q}_\ell$-algebra, we may view χ_ℓ as taking values in $(K\otimes\mathbb{Q}_\ell)^*$. Choosing $d = N$, we may use φ_N to construct the character of G

$$G \xrightarrow{\varphi_N} (\mathbb{Z}/N\mathbb{Z})^* \xrightarrow{\varepsilon} K^*$$

associated to the Dirichlet character ε. One calls this new character ε as well. Since $K\otimes\mathbb{Q}_\ell$ contains K, we may view ε as well as taking values in $(K\otimes\mathbb{Q}_\ell)^*$. The product $\varepsilon\chi_\ell^{k-1}$ is then unramified at p if $p \nmid \ell N$ and maps a Frobenius F_p for p to $\varepsilon(p)p^{k-1}$.

Proposition (2.2). We have

$$\det \rho_\ell = \varepsilon\chi_\ell^{k-1}.$$

Proof. Let ψ be the quotient of these two characters, namely $(\det\rho_\ell)(\varepsilon\chi_\ell^{k-1})^{-1}$.

Then ψ is unramified outside ℓN and takes the value 1 on Frobenius elements for all primes $p \nmid \ell N$. It is thus identically 1 by the Čebotarev density theorem [19, Ch. I, §2.2], which asserts that Frobenius elements for such primes map to a dense subset of the image of ψ.

A Remark on Eisenstein Series. For a brief moment, we permit the operators T_p and R_d to act on the space of all holomorphic modular forms of weight k on $\Gamma_1(N)$. Suppose that f is a non-zero element of this space, not necessarily a cusp form, which satisfies the equations

$$f|R_d = \varepsilon(d)f \quad (d, N) = 1$$

$$f|T_p = c_p \cdot f \quad p \nmid N.$$

The field K_f generated by the c_p and the $\varepsilon(d)$ is still finite over $\mathbb{Q}$, and given K and ℓ as in (2.1) we may wish to construct a representation ρ_ℓ. For this we may assume that f is either a cusp form or an Eisenstein series, since the space of all modular forms is the direct sum of the space of cusp forms and the space of Eisenstein series, with both spaces stable under the operators T_p and R_d. Since we know how to construct ρ_ℓ when f is a cusp form, we may assume that f is an Eisenstein series.

In that case there are characters

$$\varepsilon_1, \varepsilon_2 : (\mathbb{Z}/N\mathbb{Z})^* \to \mathbb{C}^*$$

whose product is ε, such that

$$c_p = \varepsilon_1(p)p^{k-1} + \varepsilon_2(p)$$

for all $p \nmid N$ [7, p. 690]. Under the assumption that K contains the values of the ε_i, we may thus construct ρ_ℓ as the direct sum

$$\varepsilon_1 \chi_\ell^{k-1} \oplus \varepsilon_2,$$

with the two characters being regarded as 1-dimensional representations of G

over $K \otimes \mathbb{Q}_\ell$. Establishing this assumption amounts to proving that any automorphism σ of $\mathbb{C}$ which fixes the c_p and the $\varepsilon(d)$ also fixes the values of ε_1 and ε_2.

This is easy to prove if $k > 1$. Indeed, for each prime p we have the equation

$$\sigma(\varepsilon_1(p))p^{k-1} + \sigma(\varepsilon_2(p)) = \varepsilon_1(p)p^{k-1} + \varepsilon_2(p).$$

The character values are roots of 1, and $p^{k-1} \to \infty$ as $p \to \infty$. Hence for p large we must have $\sigma(\varepsilon_1(p)) = \varepsilon_1(p)$ and $\sigma(\varepsilon_2(p)) = \varepsilon_2(p)$. By Dirichlet's theorem on arithmetic progressions, this implies that σ fixes all values of ε_1 and ε_2.

Now suppose that $k = 1$. Then ε is an odd character because k is odd. Hence ε_1 is distinct from any conjugate of ε_2. Thus from the equation

$$\sigma(\varepsilon_1) + \sigma(\varepsilon_2) = \varepsilon_1 + \varepsilon_2$$

we may deduce $\sigma(\varepsilon_1) = \varepsilon_1$, $\sigma(\varepsilon_2) = \varepsilon_2$ by the theorem on linear independence of characters. This proves what is wanted and completes the remark.

Simplicity and Uniqueness.

We return to the space of cusp forms to prove the uniqueness of the representations ρ_ℓ whose existence is asserted by (2.1). The uniqueness is an easy corollary of the following result.

Theorem (2.3). Let λ be a prime of K. Then ρ_λ is a simple K_λ-representation of G.

Proof. (cf. [5, (8.7)]). Suppose that ρ_λ is reducible over K_λ. Replacing it by an isomorphic representation, we may represent it matrically as

$$\begin{pmatrix} \varphi_1 & * \\ 0 & \varphi_2 \end{pmatrix},$$

where the φ_i are characters of G with values in K_λ^*. Let r be the representation $\varphi_1 \oplus \varphi_2$ of G. It is abelian and semisimple and has the same trace and determinant as ρ_λ. (It is the "semisimplification" of ρ_λ.) In particular, we have

$$\mathrm{tr}\, r(F_p), \det r(F_p) \in K$$

if $F_p \in G$ is a Frobenius element for a prime $p \nmid \ell N$.

Let ℓ be the residue characteristic of λ, and let $I_\ell \subset G$ be an inertia group for ℓ. By results of Serre and Lang (see [19, Ch. III]), each character φ_i may be written as an integral power of χ_ℓ, say $\chi_\ell^{n_i}$, on an open subgroup of I_ℓ. (The theorem of Lang-Serre says that r is <u>locally algebraic</u>, being a semisimple "rational" abelian λ-adic representation of the Galois group of $\mathbb{Q}$.)

On the other hand, let $p \neq \ell$ be a prime, and let I_p be an inertia group for p. Then the two characters φ_i are trivial on some open subgroup of I_p [23, p. 515]. If $p \nmid N$, this subgroup may be taken to be I_p itself, since ρ_λ is unramified outside ℓN. By class field theory, we now see that the two characters

$$\varepsilon_i = \varphi_i \chi_\ell^{-n_i}$$

are characters of finite order which are unramified outside ℓN.

Regarding them as Dirichlet characters, we may write

$$c_p = \mathrm{tr}\, r(F_p) = \varepsilon_1(p)p^{n_1} + \varepsilon_2(p)p^{n_2}$$

for all $p \nmid \ell N$. We have

$$n_1 + n_2 = k-1$$

$$\varepsilon_1 \varepsilon_2 = \varepsilon,$$

as a consequence of the equation $\varphi_1 \varphi_2 = \varepsilon \chi_\ell^{k-1}$. (The ε_i are of finite order, whereas χ_ℓ is not.) We will now conclude by showing that these formulas contradict known facts about the (archimedean) size of the c_p.

We recall the Petersson-conjecture estimate

$$|c_p| \leq 2p^{(k-1)/2}$$

(the Rankin estimate would suffice as well); it is clearly incompatible with our equation for the c_p unless $n_1, n_2 < k/2$. Since the sum of these integers is $k-1$, we must have

$$n_1 = n_2 = (k-1)/2,$$

which implies that k is odd. This implies that ε is an odd character, so that $\varepsilon_1 \neq \varepsilon_2$.

Using the non-triviality of $\varepsilon_1\varepsilon_2^{-1}$, we get that

$$\sum_{p \nmid \ell N} |c_p|^2 p^{-s} = -2\log(s-k) + O(1)$$

as $s \to k+$. However, Rankin's method [5, (5.1)] gives instead the estimate $-\log(s-k) + O(1)$ as $s \to k+$; this is a contradiction.

Corollary (2.4). Let ℓ be a prime. The $K \otimes \mathbb{Q}_\ell$-representation ρ_ℓ is unique, up to isomorphism.

Proof. We must show that any two representations ρ_ℓ, ρ'_ℓ as in (2.1) are $K \otimes \mathbb{Q}_\ell$-isomorphic. Choose $\lambda | \ell$, and let ρ_λ, ρ'_λ be the corresponding λ-adic representations. By the theorem above, these representations are simple, hence semisimple. They have identical traces on Frobenius elements, so by the Čebotarev density theorem, they have the same trace. As is well known, a semi-simple representation in characteristic 0 is determined by its trace. Hence ρ_λ and ρ'_λ are K_λ-isomorphic. Since this is true for each $\lambda | \ell$, ρ_ℓ and ρ'_ℓ are isomorphic.

§3. Modular Forms with "Complex Multiplication" and the Field K_f.

We recall an easy theorem in [5]. Let $f \in S$ be an eigenform, and let $f' \in S_{k'}(\Gamma_1(N'))$ be an eigenform of possibly different weight and level from that of f. Let $(\{c_p\},\varepsilon)$ and $(\{c'_p\},\varepsilon')$ be their respective eigenvalue "packages."

Theorem ([5, (6.3)]). Suppose that $c_p = c'_p$ for each p in a set of primes of density 1. Then $k = k'$, $c_p = c'_p$ for all $p \nmid NN'$, and $\varepsilon(d) = \varepsilon'(d)$ for all d prime to NN'.

Proof. Choose a number field K large enough to contain all the eigenvalues of the two forms, and pick a prime λ of K. Let ρ, ρ' be the λ-adic representations attached to f and f' respectively. By hypothesis, the traces of ρ and ρ' agree on Frobenius elements for all primes in a set of primes of density 1. By the Čebotarev density theorem, we find $\operatorname{tr}\rho = \operatorname{tr}\rho'$; this implies the assertion that $c_p = c'_p$ for all $p \nmid NN'$. It also follows easily that ρ and ρ' have equal determinants. (This follows in fact from the theorem that two representations with equal traces in characteristic 0 are isomorphic, or else we can use the formula

$$\det M = [(\operatorname{tr}(M))^2 - \operatorname{tr}(M^2)]/2$$

for the determinant of a 2×2 matrix in characteristic prime to 2.) So if ℓ is the residue characteristic of λ we have

$$\varepsilon\chi_\ell^{k-1} = \varepsilon'\chi_\ell^{k'-1}.$$

Since χ_ℓ is not of finite order we get $k = k'$, $\varepsilon = \varepsilon'$.

Corollary (3.1). The field K_f generated by the eigenvalues c_p, $\varepsilon(d)$ of f is already generated by the numbers c_p alone.

Proof. It suffices to show that any $\sigma \in \operatorname{Aut}(\mathbb{C})$ which fixes the c_p also fixes the $\varepsilon(d)$. We form the conjugate f^σ of f, which is an eigenform with eigenvalues $\sigma(c_p)$, $\sigma(\varepsilon(d))$. By the theorem, $\sigma(\varepsilon) = \varepsilon$ if $\sigma(c_p) = c_p$ for all p.

Proposition (3.2). Let f be an eigenform, and let $K = K_f$ be its field of eigenvalues. Let L be the largest totally real subfield of K. Then either K = L, or else K is a totally complex quadratic extension of L (i.e., K is a "CM field"). The first alternative holds when $\varepsilon = 1$.

Proof. For every $\sigma \in \mathrm{Aut}(\mathbb{C})$ we have

$$\sigma(c_p) = \overline{\sigma(c_p)} \cdot \sigma(\varepsilon(p)), \quad p \nmid N$$

where the bar denotes complex conjugation. This was noted in §1 for σ the identity map of $\mathbb{C}$, and it follows for an arbitrary σ by replacing f by its conjugate f^{σ}. Since K is generated by the numbers c_p, it follows from this equation that K has a "well-defined complex conjugation"; an easy classification theorem (cf. [24, Prop. 5.11]) then shows that K is either totally real or a CM field as asserted. Also, if $\varepsilon = 1$, then $\sigma(c_p)$ is its own complex conjugate for each σ; hence the field generated by the c_p is totally real.

Proposition (3.3). Suppose $\varepsilon \neq 1$. Then K is in fact a CM field unless ε is of order two and satisfies

$$\varepsilon(p)c_p = c_p$$

for all $p \nmid N$. Conversely, if $\varepsilon(p)c_p = c_p$ for all $p \nmid N$, then K is totally real.

Proof. Since K contains the values of ε, K can be totally real only if the values of ε are all real. Hence if K is totally real and ε is not the identity character, ε is of order 2. If K is totally real, the equation

$$c_p = \varepsilon(p)\overline{c_p}$$

becomes the identity asserted in the proposition. Conversely, the equation $\varepsilon(p)c_p = c_p$ implies that $\overline{c_p} = c_p$ and hence that c_p is a totally real number. If it holds for all $p \nmid N$, K is totally real.

Twisting. From now on, we shall assume for simplicity that f is a newform

$\Sigma a_n q^n$ of weight k on $\Gamma_1(N)$. Given a Dirichlet character φ, we let $f \otimes \varphi$ be

$$f \otimes \varphi = \sum_{n=1}^{\infty} \varphi(n) a_n q^n .$$

As is well known [24, Prop. 3.64], $f \otimes \varphi$ is a modular form of weight k on $\Gamma_1(ND^2)$ of Nebentypus $\varepsilon\varphi^2$ if φ is defined mod D. We have

$$(f \otimes \varphi) \mid T_p = [\varphi(p) a_p](f \otimes \varphi)$$

for $p \nmid ND$, so that $f \otimes \varphi$ is again an eigenform.

Definition. Suppose φ is not the trivial character. The form f has complex multiplication by φ if

$$\varphi(p) a_p = a_p$$

for all primes p in a set of primes of density 1.

Remarks.

1. If f has complex multiplication by φ, then by the above-mentioned theorem of [5] we have $\varepsilon\varphi^2 = \varepsilon$, and $\varphi(p) a_p = a_p$ for all $p \nmid DN$. The first equation implies that φ is a quadratic character. If its kernel in G corresponds to the quadratic field F, we say that f has complex multiplication by F.

2. It is immediate from the definition and from (3. 3) that a newform f whose Nebentypus character is not the identity, but which nevertheless has a totally real eigenvalue field K_f, is a form with complex multiplication by its own Nebentypus character.

Construction of Newforms with Complex Multiplication (Hecke, Shimura).

Let F be an imaginary quadratic field, and let $k > 1$ be an integer. Select an embedding $\sigma : F \hookrightarrow \mathbb{C}$. Let ψ be a Grössencharacter of F whose infinity type is σ^{k-1}. Suppose that $\mathfrak{m}$ is an integral ideal such that ψ is defined mod $\mathfrak{m}$, and view ψ as a homomorphism

$$(\text{fractional ideals of } F \text{ prime to } \mathfrak{m}) \to \mathbb{C}^*$$

such that

$$\psi((\alpha)) = \sigma(\alpha)^{k-1}$$

for all numbers $\alpha \in F^*$ such that $\alpha \equiv 1 \bmod^{\times} \mathfrak{m}$.

Let φ be the Dirichlet character associated to F, viewed as defined mod D, where $-D$ is the discriminant of F.

Let M be the norm of $\mathfrak{m}$, and let η be the Dirichlet character mod M given by the formula

$$a \mapsto \psi((a))/\sigma a^{k-1} \qquad (a \in \mathbb{Z}).$$

Let ε be the product $\eta\varphi$.

Define $g = \sum_{n=1}^{\infty} c_n q^n$ as the sum over ideals $\mathfrak{a}$ of F

$$g = \sum_{\substack{(\mathfrak{a},\mathfrak{m})=1 \\ \mathfrak{a} \text{ integral}}} \psi(\mathfrak{a}) q^{N\mathfrak{a}},$$

where $N\mathfrak{a}$ is the norm of an ideal $\mathfrak{a}$.

The following result is [25, Lemma 3]; see also [7, p. 717].

Theorem (3.4). The series g is a cusp form of weight k and character ε on $\Gamma_1(DM)$. If $p \nmid DM$, then

$$g | T_p = a_p g.$$

Corollary (3.5). There exists a unique newform $f = \sum_{n=1}^{\infty} a_n q^n$ of weight k, character ε, and level dividing DM such that

$$a_p = c_p$$

for all $p \nmid DM$. Furthermore, $a_p = 0$ if $\varphi(p) = -1$, so that f has complex multiplication by φ.

Proof. The first statement results from the general theory of newforms, as recalled in §1. For the second, one notes from the definition of g that $c_p = 0$ if no ideal of F has norm p. Also, if $a_p = 0$ when $\varphi(p) = -1$, then $a_p = \varphi(p)a_p$ for almost all p, which means that f has complex multiplication by φ.

Remark (3.5). Shimura [26, p. 138] has pointed out that g is a newform if $\mathfrak{m}$ is the conductor of ψ.

Example (Shimura). Let F be one of the imaginary quadratic fields with class number 1 and with unit group $\{\pm 1\}$. Suppose that k is odd. Define a character ψ on all non-zero fractional ideals by the rule

$$(\alpha) \mapsto \sigma\alpha^{k-1}.$$

Then ψ is a Grössencharacter as above with $\mathfrak{m} = (1)$, and in particular, we have $\eta = 1$. Hence the form f as above has Nebentypus $\varepsilon = \varphi$, so it is a form with complex multiplication by its own Nebentypus. By (3.3), this implies that the field K of eigenvalues for f is totally real. In fact, we see from the construction that the eigenvalues a_p are rational integers, so that $K = \mathbb{Q}$.

§4. Study of Complex Multiplication via λ-adic Representations.

Let f be a newform of weight k, Nebentypus ε, and level N. We wish to study possible complex multiplication of f via the ℓ-adic representations attached to f. Pick a number field $K \subset \mathbb{C}$ which contains the eigenvalues a_p (and hence the numbers $\varepsilon(d)$) for f. Let ℓ be a prime number, and let $\lambda | \ell$ be a prime of K. We proved in §2 that the representation ρ_λ attached to f is K_λ-irreducible.

Proposition (4.1). The image G_λ of ρ_λ is non-abelian.

Proof. Let $c \in G$ be a complex conjugation. Since c is of order 2, and since

$$\det\rho_\lambda(c) = \chi_\ell(c)^{k-1}\varepsilon(c) = (-1)^{k-1}\varepsilon(-1) = -1,$$

the eigenvalues of $\rho_\lambda(c)$ are $+1$, -1. These are distinct elements of K_λ^*, so the elements of $GL(2,K_\lambda)$ which commute with $\rho_\lambda(c)$ form an abelian diagonalizable subgroup T of $GL(2,K_\lambda)$. Because ρ_λ is irreducible, $G_\lambda \not\subseteq T$. Hence some element of G_λ does not commute with $\rho_\lambda(c)$, and in particular G_λ is non-abelian.

Proposition (4.2). Let H be an open subgroup of G. Then the restriction of ρ_λ to H is semisimple.

Proof. The restriction of a semisimple representation of a group to a subgroup of finite index of that group is again semisimple.

For the rest of this paper we will operate under the (tacit) assumption that $k > 1$. The assumption is certainly necessary because the theorems below are not true for $k = 1$. When $k = 1$, the groups G_λ are all (by construction) the finite image of the complex representation of G associated to f by Deligne-Serre [5]. This representation is studied in [22].

Theorem (4.3) (Serre). The image $\overline{G}_\lambda$ of G_λ in $PGL(2,K_\lambda)$ is infinite.

Proof (abstracted from a letter of Serre to the author). Suppose to the contrary that $\overline{G}_\lambda$ is finite. By the Čebotarev density theorem, infinitely many primes $p \nmid \ell N$ satisfy $\varepsilon(p) = 1$ and split in the extension of $\mathbb{Q}$ corresponding to $\overline{G}_\lambda$. Such p have Frobenius elements in G which map to scalars in $GL(2,K_\lambda)$; hence such p satisfy

$$a_p^2 = 4p^{k-1}\varepsilon(p) = 4p^{k-1}.$$

If $k-1$ is odd, this is already a contradiction: the above equation implies that infinitely many primes p become squares in K^*, contrary to the fact that only a finite number of primes can ramify in a given number field. Hence we may write $k = 2m+1$ with $m \geq 1$.

Let φ be the representation $\rho_\lambda \otimes \chi_\ell^{-m}$ of G with values in $GL(2, K_\lambda)$. The image of φ in $PGL(2, K_\lambda)$ is the same as that of ρ_λ, namely $\overline{G}_\lambda$ (a finite group). Also, the determinant of φ is a character of finite order, namely ε. This implies that the image of φ is finite. (Only the matrices $\pm I$ have determinant 1 and map to the identity in $PGL(2, K_\lambda)$.)

Let $L_\rho(s)$ and $L_\varphi(s)$ be the usual L-functions attached to the representations ρ_λ and φ (cf. [19, Ch. I, §2.5]), defined as Euler products over primes $p \nmid \ell N$. Aside from possible missing Euler factors, L_ρ is the Dirichlet series

$$\Sigma a_n n^{-s}$$

for the newform f and L_φ is an Artin L-series. Hence each L-series has a functional equation of a known type. But the relation $\rho_\lambda = \chi_\ell^m \otimes \varphi$ implies an identity

$$L_\varphi(s-m) = L_\rho(s).$$

Thus $L_\rho(s)$ has <u>two</u> functional equations, and these turn out to be incompatible, giving the desired contradiction.

First, from the modular point of view (§1), $L_\rho(s)$ has a functional equation with Γ-factor $(2\pi)^{-s}\Gamma(s)$ and symmetry $s \mapsto k-s$. From the Artin point of view, however, there is also for $L_\rho(s)$ a functional equation with Γ-factor $(2\pi)^{-s}\Gamma(s-m)$ and the same symmetry. Remembering the possible missing Euler factors, we find an identity

$$\frac{\Gamma(s)\Gamma(m+1-s)}{\Gamma(s-m)\Gamma(k-s)} = c \cdot A^s \prod_{i=1}^{t} (1-\alpha_i p_i^{-s})^{n_i},$$

where c and A are non-zero complex numbers, the n_i are integers, the α_i are complex numbers, and the p_i are primes. One sees easily that this is impossible by considering the zeros and poles of the two sides of the equation.

<u>Proposition</u> (4.4). Either $\rho_\lambda(H)$ is an irreducible non-abelian group for each

open subgroup H of G, or else there exists an open subgroup H of index 2 in G with the following property: If H' is an open subgroup of G, $\rho_\lambda(H')$ is abelian if and only if H contains H'. In the latter case, the quadratic field F corresponding to H is unramified outside ℓN, and f has complex multiplication by F. Conversely, if f has complex multiplication by a quadratic field F, then $\rho_\lambda(\mathrm{Gal}(\overline{\mathbb{Q}}/F))$ is abelian.

Proof. The first statement is an easy consequence (see [14, (2.3)]) of the fact that G_λ is non-abelian and semisimple and has an infinite image in $PGL(2, K_\lambda)$. One finds that if G_λ has an open abelian subgroup, then there is a Cartan subgroup $\mathcal{C}$ of $GL(2, K_\lambda)$ whose normalizer $\mathcal{N}$ contains G_λ. The intersection $G_\lambda \cap \mathcal{C}$ has index 2 in G_λ and its preimage in G is a closed subgroup H of index 2 on which ρ_λ is non-abelian. A subgroup H' of G is contained in H if and only if its image in $GL(2, K_\lambda)$ is contained in $\mathcal{C}$. Also, the character

$$\varphi : G \to G/H \cong \mathcal{N}/\mathcal{C} \simeq \{\pm 1\}$$

is unramified outside ℓN because ρ_λ is. Now the complement $\mathcal{N} - \mathcal{C}$ consists of matrices with trace 0. Hence if $\varphi(g) = -1$, then $\mathrm{tr}\,\rho_\lambda(g) = 0$. If we think of φ as a Dirichlet character, we may thus write

$$a_p = \mathrm{tr}\ \rho_\lambda(F_p) = 0$$

for all $p \nmid \ell N$ such that $\varphi(p) = -1$. Said another way, f has complex multiplication by φ, i.e., by F.

Now, conversely, assume that f has complex multiplication by a quadratic character φ. Let $H \subset G$ be the kernel of φ, and let F be the corresponding quadratic field. We wish to show that $\rho_\lambda(H)$ is abelian.

The fact that f has multiplication by φ shows that $\rho_\lambda \otimes \varphi$ and ρ_λ have identical traces on Frobenius elements for almost all primes p. Hence, by the Čebotarev theorem again, and because of the semisimplicity of the representations, the two representations are isomorphic. In concrete terms, this means that there is a matrix $M \in GL(2, K_\lambda)$ such that

$$M\rho_\lambda(g)M^{-1} = \varphi(g)\rho_\lambda(g)$$

for all $g \in G$. In particular, $\rho_\lambda(H)$ is contained in the commutant of M. Let $g \in G$ be an element not in H, and let $T = \rho_\lambda(g)$. Then

$$MTM^{-1} = -T.$$

This equation evidently shows that M is not a scalar matrix, and it also implies that M is semisimple. Hence the commutant of M is abelian, and this implies what we wanted.

Theorem (4.5). Suppose that there exists an open subgroup H of G of index 2 in G such that $\rho_\lambda(H)$ is abelian. Then $\rho_{\lambda'}|_H$ is abelian for all primes λ' of K. Also, H corresponds to a quadratic field F which is (i) unramified at all (finite) primes away from N, but which is (ii) ramified at infinity (i.e., imaginary). Further, f is obtained from a Grössencharacter ψ as in (3.5).

Proof. The first assertion follows from the above proposition, because the statement that f has complex multiplication by a quadratic field does not involve a prime λ of K. Also, a field F of complex multiplication is unramified away from N, because it is unramified away from ℓN, where ℓ is an arbitrary prime. However, to prove that F is imaginary and to prove that f comes from a Grössencharacter of F seem to be less elementary.

Let us suppose that f has complex multiplication by a quadratic field F, and let H be the corresponding subgroup of G. We know that $\rho_\lambda(H)$ is abelian and semisimple for each prime λ of K. But further, by theorems in [19], $\rho_\lambda|_H$ is locally algebraic for each prime λ of K, so that there is a map of K-algebraic groups

$$r : S_{\mathfrak{m}/K} \to GL(2, K)$$

giving rise to the system of representations $(\rho_\lambda|_H)$. Here $\mathfrak{m}$ is an integral ideal of F, and $S_{\mathfrak{m}}$ is the $\mathbb{Q}$-algebraic group constructed for F and $\mathfrak{m}$ as in [19, 21]. In fact, since $S_{\mathfrak{m}}$ is of multiplicative type, r is described by a K-rational pair of elements of the character group of $S_{\mathfrak{m}}$,

$$\mathrm{Hom}(S_{\mathfrak{m}/\overline{K}}, \mathbb{G}_{m/\overline{K}}).$$

(See [21] for a down-to-earth description of the elements of this group. Note that the two characters corresponding to r are not necessarily each defined over K, but they form a K-rational set.)

Now, for definiteness, choose $\overline{K}$ to be the algebraic closure of K in $\mathbb{C}$. The two characters of $S_{\mathfrak{m}}$ in question correspond to two type-A_0 Grössen-characters ψ_1 and ψ_2 of F with values in $\overline{K}^* \hookrightarrow \mathbb{C}^*$ and conductor dividing $\mathfrak{m}$. They are interchanged by the ('internal') conjugation of F over $\mathbb{Q}$. (For each prime λ of K the representation $\rho_\lambda|_H$ corresponds to two $\overline{K}_\lambda^*$-valued characters of H which are interchanged by conjugation (in G) by an element of G not in H. This conjugation takes the Frobenius elements in H for a prime ideal $\mathfrak{P}$ of F to the Frobenius elements for the conjugate ideal $\overline{\mathfrak{P}}$.) Their relation with the $\rho_\lambda|_H$ is that

$$\psi_1\psi_2 = \varepsilon \cdot (\mathrm{Norm})^{k-1}$$

(where Norm is the norm Grössencharacter of F) and that for each λ of K and any prime $\mathfrak{P}$ of F not dividing $(N\lambda)MN$ we have

$$\mathrm{tr}\,\rho_\lambda(F_{\mathfrak{P}}) = \psi_1(\mathfrak{P}) + \psi_2(\mathfrak{P}),$$

where $F_{\mathfrak{P}} \in H$ is a Frobenius for the prime $\mathfrak{P}$.

Now, using the assumption that $k > 1$, let us prove that F is complex. If not, each ψ_i be written as an integral power of the norm character times a character of finite order, say

$$\psi_i = (\mathrm{Norm})^{n_i}\varepsilon_i.$$

(See [19, Ch. II, §3.3].) This gives

$$\operatorname{tr}\rho_\lambda(F_{\mathfrak{P}}) = \varepsilon_1(\mathfrak{P})N\mathfrak{P}^{n_1} + \varepsilon_2(\mathfrak{P})N\mathfrak{P}^{n_2}$$

for all but finitely many $\mathfrak{P}$ of F. If $\mathfrak{P}|p$ is a split prime of F, the left-hand side is a_p. Hence, taking $\mathfrak{P}$ split and of large norm, we find from the archimedian estimates that $n_1 = n_2 = (k-1)/2$, as in §2. Then, over $\overline{K}_\lambda$, $\rho_\lambda|_H$ takes the shape

$$\begin{pmatrix} \varepsilon_1 & 0 \\ 0 & \varepsilon_2 \end{pmatrix} \otimes \chi_\ell^{\frac{k-1}{2}}.$$

It therefore has a finite image in $PGL(2, K_\lambda)$. This contradicts (4.3).

We continue our analysis, having established that F is imaginary. Let σ, τ be the distinct embeddings of F into $\overline{K} \subset \mathbb{C}$, and let the infinity type of ψ_1 be $\sigma^n\tau^m$. We know that the infinity type of ψ_2 is $\sigma^m\tau^n$ (because the two characters are permuted by conjugation) and that $n+m = k-1$ (because $\psi_1\psi_2 = \varepsilon(\mathrm{Norm})^{k-1}$). The above argument shows that $m \neq n$. Without loss of generality assume $n > m$. Let $\psi = \psi_1 \cdot (\mathrm{Norm})^{-m}$, so that the infinity type of ψ is σ^{n-m}. We have

$$\prod_{p\nmid NM} (1-a_p p^{-s}+\varepsilon(p)p^{k-1-2s})^{-1} = \prod_{\mathfrak{P}\nmid NM} (1-\psi_1(\mathfrak{P})N\mathfrak{P}^{-s})^{-1}$$

$$= \prod_{\mathfrak{P}\nmid NM} (1-\psi(\mathfrak{P})N\mathfrak{P}^{m-s})^{-1},$$

as can be checked p by p. Thus if g is the modular form of weight n-m+1 associated to ψ by the construction of §3, we have

$$L_f^*(s) = L_g^*(s-m),$$

where the * refers to the deletion of a finite number of Euler factors. By the argument of Serre given above (incompatibility of functional equations), we must have $m = 0$. Hence f is the newform constructed from $\psi = \psi_1$ by the process of §3.

§5. Twisting and ℓ-adic Lie Algebras.

Let f be a newform of weight k and level N. The image G_ℓ of the ℓ-adic representation attached to f is an ℓ-adic Lie group, being a closed subgroup of some linear group $GL(2, K \otimes \mathbb{Q}_\ell)$. Its Lie algebra $\mathfrak{g}_\ell$ is then a $\mathbb{Q}_\ell$-subalgebra of the corresponding Lie algebra $\mathfrak{gl}(2, K \otimes \mathbb{Q}_\ell)$ of 2×2 matrices over $K \otimes \mathbb{Q}_\ell$. In a previous paper [14], we determined $\mathfrak{g}_\ell$ in the special case of level $N = 1$, i.e., for eigenforms on $SL_2\mathbb{Z}$. As we shall see, the general case is complicated by "twisting."

To simplify matters, we will make two assumptions in this §. First of all, we again assume that $k > 1$, and secondly, we assume that f does not have complex multiplication. These assumptions are not too serious since the algebras $\mathfrak{g}_\ell$ are all 0 in the case of weight 1 (the representations ρ_ℓ have finite images) and are abelian if f has complex multiplication. (In fact, from (4.5) and results of Serre, one sees that $\mathfrak{g}_\ell = F \otimes \mathbb{Q}_\ell$, where F is the field of multiplication.)

We let K be the number field generated by the eigenvalues $a_p (p \nmid N)$, and we let L be the largest totally real subfield of K. By (3.1), K contains the values of the Nebentypus character ε of f. Also, since f does not have complex multiplication, $K = L$ <u>if and only if</u> ε is trivial. Otherwise, K is a CM field, and is quadratic over L.

Let ℓ be a prime, and let

$$\rho_\ell : G \to GL(2, K \otimes \mathbb{Q}_\ell)$$

be the ℓ-adic representation attached to f. The determinant of ρ_ℓ is $\varepsilon\chi_\ell^{k-1}$, so $\det \rho_\ell$ is $\mathbb{Q}_\ell^*$-valued on the kernel H of ε. Since H is open in G, $\mathfrak{g}_\ell$ is contained in the algebra

$$\mathfrak{b}_\ell = \{u \in \mathfrak{gl}(2, K \otimes \mathbb{Q}_\ell) \mid \operatorname{tr} u \in \mathbb{Q}_\ell\}.$$

This inclusion is an equality when $N = 1$ [14].

There is another constraint on $\mathcal{g}_\ell$ in the case where ε is non-trivial. This arises from the equation

$$a_p = \bar{a}_p \, \varepsilon(p)$$

of §1, which implies that $a_p \in L$ for all p such that $\varepsilon(p) = 1$.

<u>Proposition</u> (5.1). If H is the kernel of ε, then

$$\operatorname{tr} \rho_\ell(H) \subsetneq L \otimes \mathbb{Q}_\ell .$$

<u>Proof.</u> The Frobenii $F_p \in G$ for primes p such that $\varepsilon(p) = 1$ all lie in H, and by the Čebotarev density theorem their images in $GL(2, K \otimes \mathbb{Q}_\ell)$ are dense in $\rho_\ell(H)$. By the remark above, their traces a_p all lie in L. Hence the result follows by continuity.

<u>Corollary</u> (5.2). Suppose that k is even. Then, after replacing ρ_ℓ by an isomorphic $K \otimes \mathbb{Q}_\ell$-representation of G, we have

$$\rho_\ell(H) \subsetneq GL(2, L \otimes \mathbb{Q}_\ell) \subsetneq GL(2, K \otimes \mathbb{Q}_\ell).$$

<u>Proof.</u> Let $\lambda | \ell$ be a prime of K, and let L_λ (resp. K_λ) be the completion of L (resp. K) at λ. We contend that (up to K_λ-isomorphism) the representation ρ_λ satisfies

$$\rho_\lambda(H) \subsetneq GL(2, L_\lambda) \subsetneq GL(2, K_\lambda).$$

In fact, this is the main point: The corollary easily follows from this claim because of the theorem that two semisimple representations with the same trace are isomorphic.

To prove the contention, we have to overcome a Schur-index obstruction to defining $\rho_\lambda |_H$ over L_λ. But H contains the complex conjugations in G, since ε is an even character (it has the same parity as k). Hence $\rho_\lambda(H)$ contains the images under ρ_λ of such elements, each of which is a matrix with eigenvalues $+1, -1$. The desired statement then results from the following lemma, which may be proved by direct computation (but see [16, IXa]).

Lemma (5.3). Let $E \supset F$ be fields, and let

$$\rho : G \to GL(2, E)$$

be a semisimple representation of an arbitrary group G. Suppose that $\mathrm{tr}\,\rho(G) \subseteq F$ and that $\rho(G)$ contains a matrix whose eigenvalues are distinct elements of F. Then ρ may be defined over F, in the sense that some representation E-isomorphic to ρ takes values in $GL(2, F)$.

Corollary (5.4). Suppose that k is even, and let $\mathfrak{a}$ be the $\mathbb{Q}$-Lie algebra

$$\{u \in \mathfrak{gl}(2, L) \mid \mathrm{tr}\, u \in \mathbb{Q}\}.$$

Then

$$\mathfrak{g}_\ell \subseteq \mathfrak{a}_\ell = \mathfrak{a} \otimes \mathbb{Q}_\ell .$$

Proof. This follows from the above corollary and the previously-noted fact that $\mathrm{tr}\, \mathfrak{g}_\ell \subseteq \mathbb{Q}_\ell$.

To gain further insight into what is happening, and to investigate the situation when k is odd, we introduce the action of the "W-operator" on the representation space for ρ_ℓ. This operator is seen in the construction of ρ_ℓ (as was mentioned in §2), and we shall list axiomatically its properties as developed during [8].

In construction the representations ρ_ℓ, one begins with a certain vector space V of dimension 2 over K such that for each prime ℓ, G acts $K \otimes \mathbb{Q}_\ell$-linearly on $V_\ell = V \otimes \mathbb{Q}_\ell$. The space V is in particular an L vector space, of dimension 4 if $L \neq K$. Already on V there is an L-linear operator W which satisfies

$$W^2 = (-N)^{k-2}$$

and the commutation relation

$$WtW^{-1} = \bar{t}$$

for $t \in K$, where the bar is, as usual, conjugation of K over L.

For each prime ℓ, the actions of G and W on V_ℓ commute according to the formula

$$W\rho_\ell(g) = \rho_\ell(g)\varepsilon^{-1}(g)W = \rho_\ell(g)W\varepsilon(g),$$

and in particular the actions of W and H commute with each other.

Let $A \subset \mathrm{End}(V)$ be the L-algebra generated by W and the elements of K. Then

$$\rho_\ell(H) \subseteq \mathrm{Aut}_{A \otimes \mathbb{Q}_\ell}(V \otimes \mathbb{Q}_\ell).$$

Let us analyze this group of automorphisms in the case which is not yet treated, that where k is odd.

For this, choose $X \in K$ such that $X^2 = \alpha$ is a totally negative element of L. We have $\bar{X} = -X$. Set $Z = XW$, and let $\beta = -N^{k-2}$. Then:

$$WX = \bar{X}W = -XW = -Z,$$
$$W^2 = \beta,$$
$$X^2 = \alpha,$$
$$Z^2 = -\beta\alpha,$$
$$XZ = -ZX = \alpha W,$$
$$WZ = -ZW = -\beta X.$$

In other words, A is the quaternion algebra over L associated to α, β (Bourbaki, Alg. III., p. 18). It is a division algebra, and in fact a totally definite quaternion algebra, because α and β are totally negative.

Since $\dim_L V = 4 = \dim_L A$, the (left) A-module V is free of rank 1 over A: V is isomorphic to A acting on itself by left multiplication. It is then immediate that if we identify V with A then $\mathrm{End}_A V$ is the set of endomorphisms of V gotten by letting elements of A act on $V = A$ on the <u>right</u>. Take $\{1, W\}$

as a K-basis for V. Then we find concretely that right multiplication by t+uW (with $t, u \in K$) is the matrix

$$M_{t,u} = \begin{pmatrix} t & \bar{u}\beta \\ u & \bar{t} \end{pmatrix}$$

and that $B = \mathrm{End}_A V$ is the subalgebra of $\mathrm{End}_K V$ consisting of all such matrices. Remembering that A is isomorphic to its own opposite algebra under the standard quaternion involution

$$t+uW \mapsto \bar{t}-uW,$$

we find that

$$t+uW \mapsto M_{\bar{t},-u}$$

induces an isomorphism $A \xrightarrow{\sim} B$. Any matrix $M_{t,u}$ has trace and determinant in L; these numbers are respectively the (reduced) trace and norm of t+uW.

We summarize all this in the following

Theorem (5.5). Let $A \subset \mathrm{End}_L(V)$ be the algebra generated over L by the operator W and the elements of K. Then A is a totally definite quaternion algebra over L, as is its commutant

$$B = \mathrm{End}_A V.$$

For each prime ℓ, we have

$$\rho_\ell(H) \subseteq \mathrm{Aut}_{A\otimes\mathbb{Q}_\ell}(V_\ell) = (B\otimes\mathbb{Q}_\ell)^*.$$

Furthermore, $\rho_\ell(H)$ is contained in the group of elements of $(B\otimes\mathbb{Q}_\ell)^*$ whose reduced norms to $(L\otimes\mathbb{Q}_\ell)^*$ lie in $\mathbb{Q}_\ell^*$.

In analogy with the situation when k is even, we let $\mathfrak{a}$ be the $\mathbb{Q}$-Lie algebra

$$\{u \in B \mid \mathrm{tr}\, u \in \mathbb{Q}\} \subset \mathfrak{gl}(2,K).$$

Corollary (5.6). For each prime ℓ, we have

$$\mathfrak{g}_\ell \subseteq \mathfrak{a}_\ell = \mathfrak{a}\otimes\mathbb{Q}_\ell.$$

The algebra $\mathfrak{a}$ in the case of odd weight is a form of the algebra which arises in the case where k is even, namely,

$$\{u \in \mathfrak{gl}(2,L) \mid \operatorname{tr} u \in \mathbb{Q}\}.$$

(In other words, the two become equal over $\overline{\mathbb{Q}}$.) Both in the case where k is even and in the case where k is odd we have defined a Lie algebra $\mathfrak{a} \subset \mathfrak{gl}(2,K)$, depending on f, such that

$$\mathfrak{g}_\ell \subseteq \mathfrak{a} \otimes \mathbb{Q}_\ell$$

for all primes ℓ. We now address the question of when this inclusion may be strict.

Extra Twisting. Let us say that a newform f admits an extra twist if f has complex multiplication or if there exists a $\tau \in \operatorname{Aut}(\mathbb{C})$ which does not pointwise fix L, together with a (continuous) character

$$\varphi: G \to \mathbb{C}^*$$

unramified outside N, such that

$$a_p = \tau(a_p) \cdot \varphi(p)$$

for all primes p in a set of primes of density 1.

Examples. Several examples of forms with extra twisting and without complex multiplication have been given in weight k=2 and with Nebentypus $\varepsilon = 1$. The first two examples were noted by Doi and Yamauchi [6], and others have recently been found by Birch (unpublished) and by Koike [10]. The first example of [6] is a form f on $\Gamma_0(125)$ of weight 2 and Nebentypus $\varepsilon = 1$ for which

$$K = L = \mathbb{Q}(\sqrt{4+\sqrt{5}}).$$

We have $a_p = \tau(a_p)\varphi(p)$ for all $p \neq 5$, where φ is the quadratic character associated to $\mathbb{Q}(\sqrt{5})$ and where τ conjugates K over $\mathbb{Q}(\sqrt{5})$.

It would be of interest to give an a priori construction of forms with extra twisting. If the level N is divisible by a high power of a prime, these forms seem to be more the rule than the exception.

Theorem (5.7). Let f be a newform of weight $k > 1$ which does not have complex multiplication. Let ℓ be a prime. We have the strict inclusion

$$\mathfrak{g}_\ell < \mathfrak{a}_\ell$$

if and only if f admits an extra twist.

Proof. We shall merely sketch the proof, which follows the proof of [15, (4.4.10)]. By extension of scalars, we regard ρ_ℓ as the $K \otimes \overline{\mathbb{Q}}_\ell$-representation

$$G \xrightarrow{\rho_\ell} GL(2, K \otimes \mathbb{Q}_\ell) \hookrightarrow GL(2, K \otimes \overline{\mathbb{Q}}_\ell).$$

Changing bases if necessary, we may assume that ρ_ℓ maps H into $GL(2, L \otimes \overline{\mathbb{Q}}_\ell)$ and that $\mathfrak{a}_\ell \otimes \overline{\mathbb{Q}}_\ell \subset \mathfrak{gl}(2, L \otimes \overline{\mathbb{Q}}_\ell)$. For each embedding

$$\sigma : K \hookrightarrow \overline{\mathbb{Q}}_\ell,$$

we let ρ_σ be the composite

$$G \xrightarrow{\rho_\ell} GL(2, K \otimes \overline{\mathbb{Q}}_\ell) \to GL(2, \overline{\mathbb{Q}}_\ell),$$

with the latter map being induced by σ. To calculate the Lie algebra of ρ_ℓ, we may restrict ρ_ℓ to H. On H, ρ_σ equals ρ_τ if ρ_σ and ρ_τ agree on L. We note the following facts: (i) ρ_ℓ is semisimple; (ii) the determinant of ρ_ℓ is of infinite order; (iii) no representation ρ_λ (with $\lambda | \ell$) becomes abelian on an open subgroup of G (because f does not have complex multiplication). By applying the analysis of [15, Ch. IV, §4] we find that $\mathfrak{g}_\ell < \mathfrak{a}_\ell$ if and only if there exist embeddings σ, τ, distinct on L, such that the two representations ρ_σ and ρ_τ of G are isomorphic on an open subgroup of G. By the (easy) fact that $\rho_\sigma(H_0)$ has commutant in $GL(2, \overline{\mathbb{Q}}_\ell)$ consisting of scalars for any open subgroup H_0 of G, we find that if $\mathfrak{g}_\ell < \mathfrak{a}_\ell$ there is a continuous character

$$\varphi : G \to \overline{\mathbb{Q}}_\ell^*$$

of finite order such that ρ_τ and $\rho_\sigma \otimes \varphi$ are isomorphic representations of G. Clearly φ is unramified outside N because the ρ's are, and we end up with the equation

$$\tau(a_p) = \sigma(a_p)\varphi(p) \qquad (*)$$

for all $p \nmid N\ell$. Conversely, if such an equation holds for all p in a set of primes of density 1, then we have $\rho_\tau \approx \rho_\sigma \otimes \varphi$ by the Čebotarev density theorem, and this implies $q_\ell < a_\ell$.

But (*) is purely algebraic in the sense that in (*) we may replace $\overline{\mathbb{Q}}_\ell$ by any algebraically closed field of characteristic 0: for instance, $\mathbb{C}$. More specifically, (*) holds for some σ, τ, φ as above if and only if we have

$$\tau(a_p) = \sigma(a_p)\varphi(p)$$

for some elements σ, τ of Aut($\mathbb{C}$) which are distinct on L and for some complex-valued character of finite order φ. Hence (*) holds if and only if f has a non-trivial twist.

<u>Correction.</u> The hypothesis to [14, (4.3)] is misstated. It should read $\ell > k$, not $\ell \nmid k$.

BIBLIOGRAPHY

[1] Atkin, A. O. L., Lehner, J.: Hecke operators on $\Gamma_0(m)$, Math. Ann., 185 (1970), 134-160.

[2] Casselman, W.: On some results of Atkin and Lehner, Math. Ann., 201 (1973), 301-314.

[3] Deligne, P.: Formes modulaires et représentations ℓ-adiques, Séminaire Bourbaki, 355, Février, 1969. Lecture Notes in Mathematics, 179, Springer, Berlin-Heidelberg-New York, 1971.

[4] Deligne, P.: La conjecture de Weil I, Publ. Math. I. H. E. S., 43 (1974), 273-307.

[5] Deligne, P., Serre, J -P.: Formes modulaires de poids 1, Ann. Scient. Ec. Norm. Sup., 4e série, 7 (1974), 507-530.

[6] Doi, K., Yamauchi, M.: On the Hecke operators for $\Gamma_0(N)$ and class fields over quadratic number fields, J. Math. Soc. Japan, 25 (1973), 629-643.

[7] Hecke, E.: Mathematische Werke, Vandenhoeck und Ruprecht, Göttingen, 1959.

[8] Katz, N.: Course at Princeton University, 1974-1975.

[9] Koike, M.: Congruences between cusp forms and linear representations of the Galois group (to appear).

[10] Koike, M.: On certain abelian varieties obtained from new forms of weight 2 on $\Gamma_0(3^4)$ and $\Gamma_0(3^5)$, Nagoya Math. J., 62 (1976), 29-39.

[11] Li, W.-C.: Newforms and functional equations, Math. Ann., 212 (1975), 285-315.

[12] Miyake, T.: On automorphic forms on GL_2 and Hecke operators, Ann. of Math., 94 (1971), 174-189.

[13] Rankin, R. A.: Contributions to the theory of Ramanujan's function $\tau(n)$ and similar arithmetical functions I, II, Proc. Camb. Phil. Soc., 35 (1939), 351-372.

[14] Ribet, K. A.: On ℓ-adic representations attached to modular forms, Inventiones Math., 28 (1975), 245-275.

[15] Ribet, K. A.: Galois action on division points of abelian varieties with many real multiplications, Am. J. Math., 98 (1976), 751-804.

[16] Schur, I.: Arithmetische Untersuchungen über endliche Gruppen linearer Substitutionen, Gesam. Abhl. I, 177-197. Springer, Berlin-Heidelberg-New York, 1973.

[17] Selberg, A: On the estimation of Fourier coefficients of modular forms, Proc. Sym. Pure Math. VIII, AMS, Providence, 1965.

[18] Serre, J -P.: Une interprétation des congruences relatives à la fonction τ de Ramanujan, Séminaire Delange-Pisot-Poitou, 1967-1968, ex. 14.

[19] Serre, J -P.: Abelian ℓ-adic Representations and Elliptic Curves, Benjamin, New York, 1968.

[20] Serre, J -P.: Congruences et formes modulaires (d'après H. P. F. Swinnerton-Dyer), Séminaire Bourbaki, 416, Juin, 1972. Lecture Notes in Mathematics, 317, Springer, Berlin-Heidelberg-New York, 1973.

[21] Serre, J -P.: Propriétés galoisiennes des points d'ordre fini des courbes elliptiques, Inventiones Math., 15 (1972), 259-331.

[22] Serre, J -P.: Lectures on modular forms of weight 1, Proceedings of the 1975 Durham Conference on Number Theory (to appear).

[23] Serre, J -P., Tate, J.: Good reduction of abelian varieties, Ann. of Math., 88 (1968), 492-517.

[24] Shimura, G.: Introduction to the Arithmetic Theory of Automorphic Functions, Publ. Math. Soc. Japan, No. 11, Tokyo-Princeton, 1971.

[25] Shimura, G.: On elliptic curves with complex multiplication as factors of the Jacobians of modular function fields, Nagoya Math. J., 43 (1971), 199-208.

[26] Shimura, G.: Class fields over real quadratic fields and Hecke operators, Ann. of Math., 95 (1972), 130-190.

[27] Shimura, G.: On the factors of the Jacobian variety of a modular function field, J. Math. Soc. Japan, 25 (1973), 523-544.

[28] Swinnerton-Dyer, H. P. F.: On ℓ-adic representations and congruences for coefficients of modular forms, International Summer School on Modular Functions, Antwerp, 1972. Lecture Notes in Mathematics, 350, Springer, Berlin-Heidelberg-New York, 1973.

Prof.K. Ribet
Department of Mathematics
Fine Hall
Princeton University
Princeton, N.J. 08540

A RESULT ON MODULAR FORMS IN CHARACTERISTIC p

Nicholas M. Katz

ABSTRACT

The action of the derivation $\theta = q\frac{d}{dq}$ on the q-expansions of modular forms in characteristic p is one of the fundamental tools in the Serre/Swinnerton-Dyer theory of mod p modular forms. In this note, we extend the basic results about this action, already known for $p \geq 5$ and level one, to arbitrary p and arbitrary prime-to-p level.

I. Review of modular forms in characteristic p

We fix an algebraically closed field K of characteristic $p > 0$, an integer $N \geq 3$ prime to p, and a primitive N'th root of unity $\zeta \in K$. The moduli problem "elliptic curves E over K-algebras with level N structure α of determinant ζ" is represented by

$$\begin{array}{c}(E_{univ}, \alpha_{univ}) \\ \downarrow \pi \\ M_N\end{array}$$

with M_N a smooth affine irreducible curve over K. In terms of the invertible sheaf on M_N

$$\underline{\omega} = \pi_* \Omega^1_{E_{univ}/M_N} ,$$

the graded ring $R^{\cdot}_N$ of (not necessarily holomorphic at the cusps) level N modular forms over K is

$$\bigoplus_{k \in \mathbb{Z}} H^0(M_N, \underline{\omega}^{\otimes k})$$

Given a K-algebra B, a test object (E,α) over B, and a nowhere-vanishing invariant differential ω on E, any element

$f \in R_N^{\cdot}$ (not necessarily homogeneous) has a value $f(E,\omega,\alpha) \in B$, and f is determined by all of its values (cf. [2]).

Over $B = K((q^{1/N}))$, we have the Tate curve Tate(q) with its "canonical" differential ω_{can} (viewing Tate(q) as $\mathbb{G}_m/q^{\mathbb{Z}}$, ω_{can} "is" dt/t from $\mathbb{G}_m$). By evaluating at the level N structures α_0 of determinant ζ on Tate(q), all of which are defined over $K((q^{1/n}))$, we obtain the q-expansions of elements $f \in R_N^{\cdot}$ at the corresponding cusps:

$$f_{\alpha_0}(q) \overset{\text{dfn}}{=\!=} f(\text{Tate}(q),\omega_{can},\alpha_0) \in K((q^{1/N})) \quad .$$

A homogeneous element $f \in R_N^k$ is uniquely determined by its weight k and by any one of its q-expansions. A form $f \in R_N^k$ is said to be holomorphic if all of its q-expansions lie in $K[[q^{1/N}]]$, and to be a cusp form if all of its q-expansions lie in $q^{1/N}K[[q^{1/N}]]$. The holomorphic forms constitute a subring $R_{N,holo}^{\cdot}$ of $R_N^{\cdot}$, and the cusp forms are a graded ideal in $R_{N,holo}^{\cdot}$.

The Hasse invariant $A \in R_{N,holo}^{p-1}$ is defined modularly as follows. Given (E,ω,α) over B, let $\eta \in H^1(E,\mathcal{O}_E)$ be the basis dual to $\omega \in H^0(E,\Omega^1_{E/B})$. The p'th power endomorphism $x \to x^p$ of $\mathcal{O}_E$ induces an endormorphism of $H^1(E,\mathcal{O}_E)$, which must carry η to a multiple of itself. So we can write

$$\eta^p = A(E,\omega,\alpha)\cdot\eta \quad \text{in} \quad H^1(E,\mathcal{O}_E),$$

for some $A(E,\omega,\alpha) \in B$, which is the value of A on (E,ω,α). <u>All</u> the q-expansions of the Hasse invariant are identically 1:

$$A_{\alpha_0}(q) = 1 \quad \text{in} \quad K((q^{1/N})) \text{ for each } \alpha_0.$$

For each level N structure α_0 on Tate(q), the corresponding q-expansion defines ring homomorphisms

$$R_N^{\cdot} \longrightarrow K((q^{1/N})), \quad R_{N,holo}^{\cdot} \longrightarrow K[[q^{1/N}]]$$

whose kernels are precisely the principal ideals $(A-1)R_N^{\cdot}$ and

$(A-1)R^{\cdot}_{N,\text{holo}}$ respectively ([4], [5]).

A form $f \in R^k_N$ is said to be of <u>exact filtration</u> k if it is not divisible by A in $R^{\cdot}_N$, or equivalently, if there is no form $f' \in R^{k'}_N$ with $k' < k$ which, at some cusp, has the same q-expansion that f does.

II. Statment of the theorem, and its corollaries

The following theorem is due to Serre and Swinnerton-Dyer ([4], [5]) in characteristic $p \geq 5$, and level $N = 1$.

<u>Theorem.</u>

(1) There exists a derivation $A\theta : R^{\cdot}_N \to R^{\cdot+p+1}_N$ which increases degrees by $p + 1$, and whose effect upon each q-expansion is $q \frac{d}{dq}$:

$$(A\theta f)_{\alpha_0}(q) = q \frac{d}{dq}(f_{\alpha_0}(q)) \text{ for each } \alpha_0 .$$

(2) If $f \in R^k_N$ has exact filtration k, and p does not divide k, then $A\theta f$ has exact filtration $k + p + 1$, and in particular $A\theta f \neq 0$.

(3) If $f \in R^{pk}_N$ and $A\theta f = 0$, then $f = g^p$ for a unique $g \in R^k_N$.

<u>Some Corollaries</u>

(1) The operator $A\theta$ maps the subring of holomorphic forms to the ideal of cusp forms. (Look at q-expansions.)

(2) If f is non-zero and holomorphic, of weight $1 \leq k \leq p - 2$, then f has exact filtration k. (For if $f = Ag$, then g is holomorphic of weight $k - (p-1) < 0$, hence $g = 0$.)

(3) If $1 \leq k \leq p - 2$, the map $A\theta : R^k_{N,\text{holo}} \to R^{k+p+1}_{N,\text{holo}}$ is injective. (This follows from (2) above and the theorem.)

(4) If f is non-zero and holomorphic of weight $p - 1$, and vanishes at some cusp, then f has exact filtration $p - 1$. (For if $f = Ag$, then g is holomorphic of weight 0, hence constant; as g vanishes at one cusp, it must be zero.)

(5) (determination of Ker($A\theta$)). If $f \in R_N^k$ has $A\theta f = 0$, then we can uniquely write $f = A^r \cdot g^p$ with $0 \leq r \leq p-1$, $r + k \equiv 0 (\mathrm{mod}\ p)$, and $g \in R_N^\ell$ with $p\ell + r(p-1) = k$. (This is proven by induction on r, the case $r = 0$ being part (3) of the theorem. If $r \neq 0$, then $k \not\equiv 0(p)$, but $A\theta f = 0$. Hence by part (2) of the theorem $f = Ah$ for some $h \in R_N^{k+1-p}$. Because f and h have the same q-expansions, we have $A\theta h = 0$, and h has lower r.)

(6) In (5) above, if f is holomorphic (resp. a cusp form, resp. invariant by a subgroup of $SL_2(Z/NZ)$), so is g (by unicity of g).

III. Beginning of the proof: defining θ and $A\theta$, and proving part (1)

The absolute Frobenius endomorphism F of M_N induces an F-linear endomorphism of $H^1_{DR}(E_{univ}/M_N)$, as follows. The pull-back $E_{univ}^{(F)}$ of E_{univ} is obtained by dividing E_{univ} by its finite flat rank p subgroup scheme Ker Fr where

$$Fr: E_{univ} \longrightarrow E_{univ}^{(F)}$$

is the relative Frobenius morphism. The desired map is Fr^*

$$\begin{array}{ccc} Fr^*: H^1_{DR}(E_{univ}^{(F)}/M_N) & \longrightarrow & H^1_{DR}(E_{univ}/M_N) \\ \uparrow \wr & & \\ H^1_{DR}(E_{univ}/M_N)^{(F)} & & \end{array}$$

Lemma 1. The image U of Fr^* is a locally free submodule of $H^1_{DR}(E_{univ}/M_N)$ of rank one, with the quotient $H^1_{DR}(E_{univ}/M_N)/U$ locally free of rank one. The open set $M_N^{Hasse} \subset M_N$ where A is invertible is the largest open set over which U splits the Hodge filtration, i.e., where $\underline{\omega} \oplus U \xrightarrow{\sim} H^1_{DR}(E_{univ}/M_N)$.

Proof. Because Fr^* kills $H^0(E_{univ}, \Omega^1 E_{univ/M_N})^{(F)}$ it factors through the quotient $H^1(E_{univ}, \mathcal{O})^{(F)}$, where it induces the inclusion map in the "conjugate filtration" short exact sequence (cf [1], 2.3)

$$0 \longrightarrow H^1(E_{univ},\mathcal{O})^{(F)} \xrightarrow{Fr^*} H^1_{DR}(E_{univ}/M_N) \longrightarrow H^0(E_{univ}, \Omega^1 E_{univ/M_N}) \longrightarrow 0.$$

This proves the first part of the lemma. To see where U splits the Hodge filtration, we can work locally on M_N. Choose a basis ω, η of H^1_{DR} adapted to the Hodge filtration, and satisfying $\langle\omega,\eta\rangle_{DR} = 1$. Then η projects to a basis of $H^1(E,\mathcal{O}_E)$ dual to ω, and so the matrix of Fr^* on H^1_{DR} is (remembering $Fr^*(\omega^{(F)}) = 0$)

$$\begin{pmatrix} 0 & B \\ 0 & A \end{pmatrix}$$

where A is the value of the Hasse invariant. Thus U is spanned by $B\omega + A\eta$, and the condition that ω and $B\omega + A\eta$ together span H^1_{DR} is precisely that A be invertible. Q.E.D.

<u>Remark</u>. According to the first part of the lemma, the functions A and B which occur in the above matrix have no common zero. This will be crucial later.

We can now define a derivation θ of $R_N[1/A]$ as follows. (Compare [2], A1.4.) Over M_N^{Hasse}, we have the decomposition

$$H^1_{DR} \simeq \underline{\omega} \oplus U \quad ,$$

which for each integer $k \geq 1$ induces a decomposition

$$\mathrm{Symm}^k H^1_{DR} \simeq \underline{\omega}^{\otimes k} \oplus (\underline{\omega}^{\otimes k-1} \otimes U) \oplus \ldots \oplus U^{\otimes k}$$

The Gauss-Manin connection

$$\nabla : H^1_{DR} \longrightarrow H^1_{DR} \otimes \Omega^1_{M_N/K}$$

induces, for each $k \geq 1$ a connection

$$\nabla : \mathrm{Symm}^k H^1_{DR} \longrightarrow (\mathrm{Symm}^k H^1_{DR}) \otimes \Omega^1_{M_N/K} \quad .$$

Using the Kodaira-Spencer isomorphism ([2], A.1.3.17)

$$\underline{\omega}^{\otimes 2} \xrightarrow{\sim} \Omega^1_{M_N/K}$$

we can define a mapping of sheaves

$$\theta : \underline{\omega}^{\otimes k} \longrightarrow \underline{\omega}^{\otimes k+2}$$

as the composite

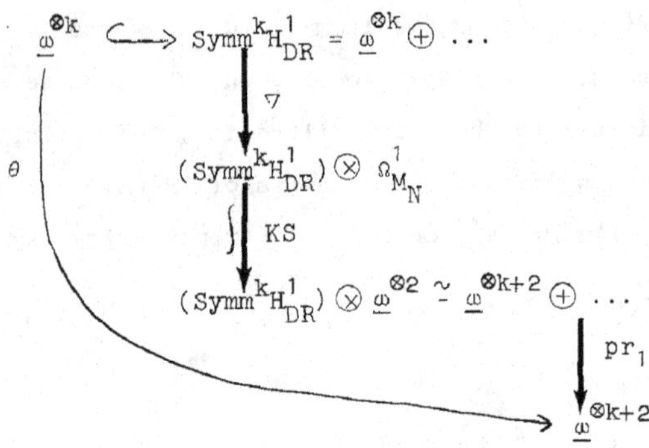

Passing to global sections over M_N^{Hasse}, we obtain, for $k \geq 1$, a map

$$\theta : H^0(M_N^{\text{Hasse}}, \underline{\omega}^{\otimes k}) \longrightarrow H^0(M_N^{\text{Hasse}}, \underline{\omega}^{\otimes k+2}) \quad .$$

Lemma 2. The effect of θ upon q-expansions is $q\frac{d}{dq}$.

Proof. Consider Tate(q) with its canonical differential $\omega_{can}^{\otimes 2}$ over $k((q^{1/N}))$ Under the Kodaira-Spencer isomorphism, ω_{can} corresponds to dq/q, the dual derivation to which is $q\frac{d}{dq}$. By the explicit calculations of ([2], A.2.2.7), U is spanned by $\nabla(q\frac{d}{dq})(\omega_{can})$. Thus given an element $f \in H^0(M_N^{\text{Hasse}}, \underline{\omega}^{\otimes k})$, its local expression as a section of $\underline{\omega}^{\otimes k}$ on (Tate(q), some α_0) is $f_{\alpha_0}(q)\cdot\omega_{can}^{\otimes k}$. Thus

$$\begin{aligned}
\nabla(f_{\alpha_0}(q)\cdot\omega_{can}^{\otimes k}) &= \nabla(q\tfrac{d}{dq})(f_{\alpha_0}(q)\cdot\omega_{can}^{\otimes k}) \cdot \frac{dq}{q} \\
&= \nabla(q\tfrac{d}{dq})(f_{\alpha_0}(q)\cdot\omega_{can}^{\otimes k}) \cdot \omega_{can}^{\otimes 2} \\
&= q\tfrac{d}{dq}(f_{\alpha_0}(q))\cdot\omega_{can}^{\otimes k+2} + k\cdot f_{\alpha_0}(q)\cdot\omega_{can}^{\otimes k+1}\cdot\nabla(q\tfrac{d}{dq})(\omega_{can}).
\end{aligned}$$

Because $\nabla(q\frac{d}{dq})(\omega_{can})$ lies in U, it follows from the definition of θ that we have $(\theta f)_{\alpha_0}(q) = q\frac{d}{dq}(f_{\alpha_0}(q))$.

Q.E.D.

Lemma 3. For $k \geq 1$, there is a unique map $A\theta : R_N^k \longrightarrow R_N^{k+p+1}$ such that the diagram below commutes

$$\begin{array}{ccccc} H^0(M_N^{Hasse}, \underline{\omega}^{\otimes k}) & \xrightarrow{\theta} H^0(M_N^{Hasse}, \underline{\omega}^{\otimes k+2}) & \xrightarrow{\times A} & H^0(M_N^{Hasse}, \underline{\omega}^{\otimes k+p+1}) \\ \cup & & & \cup \\ R_N^k = H^0(M_N, \underline{\omega}^{\otimes k}) & \xrightarrow{\qquad A\theta \qquad} & & R_N^{k+p+1} = H^0(M_N, \underline{\omega}^{\otimes k+p+1}) \end{array}$$

Proof. Again we work locally on M_N. Let ω be a local basis of $\underline{\omega}$, ξ the local basis of $\Omega^1_{M_N/K}$ corresponding to by the Kodaira-Spencer isomorphism, D the local basis of $\underline{\mathrm{Der}}_{M_N/K}$ dual to ξ, and $\omega' = \nabla(D)\omega \in H^1_{DR}$. Then $\langle\omega,\omega'\rangle_{DR} = 1$, (this characterizes D), so that ω and ω' form a basis of H^1_{DR}, adapted to the Hodge filtration, in terms of which the matrix of Fr^* is

$$\begin{pmatrix} 0 & B \\ 0 & A \end{pmatrix}$$

with $A = A(E,\omega)$. Let $u \in U$ be the basis of U over M_N^{Hasse} which is dual to ω. Then u is proportional to $B\omega + A\omega'$, and satisfies $\langle\omega,\omega'\rangle_{DR} = 1$, so that

$$u = \frac{B}{A}\omega + \omega' .$$

In terms of all this, we will compute θf for $f \in R_N^k$, and show that it has at worst a single power of A in its denominator. Locally, f is the section $f_1 \cdot \omega^{\otimes k}$ of $\underline{\omega}^{\otimes k}$, with f_1 holomorphic.

$$\begin{aligned} \nabla(f_1\omega^{\otimes k}) &= \nabla(D)(f_1\omega^{\otimes k})\cdot\xi \\ &= \nabla(D)(f_1\omega^{\otimes k})\cdot\omega^{\otimes 2} \\ &= D(f_1)\cdot\omega^{\otimes k+2} + kf_1\omega^{\otimes k+1}\cdot\omega' \\ &= D(f_1)\omega^{\otimes k+2} + kf_1\omega^{\otimes k+1}(u - \frac{B}{A}\omega) \end{aligned}$$

$$= (D(f_1) - kf_1 \cdot \tfrac{B}{A})\, \omega^{\otimes k+2} + kf_1 \omega^{\otimes k+1} \cdot u \quad .$$

Thus from the definition of θ it follows that the local expression of $\theta(f)$ is

$$\theta(f) = (D(f_1) - kf_1 \tfrac{B}{A})\, \omega^{\otimes k+2} \quad . \qquad \text{Q.E.D.}$$

We can now conclude the proof of Part (1) of the theorem. Up to now, we have only defined A on elements of $\dot{R}_N$ of positive degree. But as $\dot{R}_N$ has units which are homogeneous of positive degree (e.g., Δ), the derivation $A\theta$ extends uniquely to all of $\dot{R}_N$ by the explicit formula

$$A\theta f = \frac{A\theta(f\cdot\Delta^{pr})}{\Delta^{pr}} \quad \text{for} \quad r \gg 0 \quad .$$

The local expression for $A\theta(f)$

$$A\theta(f) = (AD(f_1)-kf_1B)\omega^{\otimes k+p+1} \quad \text{for} \quad f \in R_N^k$$

remains valid.

IV. Conclusion of the proof: Parts (2) and (3)

Suppose $f \in R_N^k$ has exact filtration k. This means that f is not divisible by A in $\dot{R}_N$, i.e., that at some zero of A, f has a lower order zero (as section of $\underline{\omega}^{\otimes k}$) than A does (as section of $\underline{\omega}^{\otimes p-1}$). (In fact, we know by Igusa [3] that A has simple zeros, so in fact f must be invertible at some zero of A. Rather surprisingly, we will not make use of this fact.)

Locally on M_N, we pick a basis ω of $\underline{\omega}$. Then f becomes $f_1 \cdot \omega^{\otimes k}$, and $A\theta(f)$ is given by

$$A\theta(f_1 \cdot \omega^{\otimes k}) = (AD(f_1)-kBf_1)$$

Suppose now that k is not divisible by p. Recall that B is invertible at all zeros of A (cf the remark following Lemma 1). Thus if $x \in M_N$ is a zero of A where $\text{ord}_x(f_1) < \text{ord}_x(A)$, we easily compute

$$\mathrm{ord}_x(AD(f_1)-kBf_1) = \mathrm{ord}_x(f_1) < \mathrm{ord}_x(A).$$

This proves Part (2) of the theorem.

To prove Part (3), let $f \in R_N^{pk}$ have $A\theta(f) = 0$. The local expression for $A\theta(f)$ gives

$$AD(f_1)\omega^{\otimes k+p+1} = 0$$

and hence $D(f_1) = 0$. Because M_N is a smooth curve over a perfect field of characteristic p, this implies that f_1 is a pth power, say $f_1 = (g_1)^p$. Thus $f_1\omega^{\otimes kp} = (g_1\omega^{\otimes k})^p$, so that f, as section of $\underline{\omega}^{\otimes kp}$, is, locally on M_N, the pth power of a (necessarily unique) section g of $\underline{\omega}^{\otimes k}$. By unicity, these local g's patch together. Q.E.D.

REFERENCES

[1] N. Katz, Algebraic solutions of differential equations, Inventiones Math. 18 (1972), 1-118.

[2] ———, P-adic properties of modular schemes and modular forms, Proceedings of the 1972 Antwerp International Summer School on Modular Functions, Springer Lecture Notes in Mathematics 350 (1973), 70-189.

[3] J. Igusa, Class number of a definite quaternion with prime discriminant, Proc. Nat'l. Acad. Sci. 44 (1958), 312-314.

[4] J. -P. Serre, Congruences et formes modulaires (d'apres H. P. F. Swinnerton-Dyer), Exposé 416, Séminaire N. Bourbaki, 1971/72, Springer Lecture Notes in Mathematics 317 (1973), 319-338.

[5] H. P. F. Swinnerton-Dyer, On ℓ-adic representations and congruences for coefficients of modular forms, Proceedings of the 1972 Antwerp International Summer School on Modular Forms, Springer Lecture Notes in Mathematics 350 (1973), 1-55.

N. Katz
Departement of Mathematics
Fine Hall
Princeton University
Princeton, N.J. 08540

ON ℓ-ADIC REPRESENTATIONS AND CONGRUENCES FOR COEFFICIENTS OF MODULAR FORMS (II)

H.P.F. Swinnerton-Dyer

1. INTRODUCTION.

Let $f = \sum \tau_k(n)q^n$ be a cusp form of weight k for the full modular group, with the properties that $\tau(1) = 1$, that every $\tau(n)$ is in $\mathbf{Z}$, and that the associated Dirichlet series has an Euler product

$$\sum \tau(n)n^{-s} = \Pi(1-\tau(p)p^{-s} + p^{k-1-2s})^{-1}. \tag{1}$$

For example, f may be the unique normalized cusp form of weight 12, 16,18,20,22 or 26; by omitting the subscript k we do not imply the traditional usage that k = 12.

It has long been known (at least in the case k = 12) that the $\tau(p)$ satisfy congruences modulo powers of certain small primes. In a previous paper [3] with the same title, which was based on joint work with Serre, it was shown that the existence and form of such congruences was closely linked to the Serre-Deligne representation theorem. By determining the structure of the graded ring of modular forms mod ℓ, we were able to establish a finite list of possible congruences for the $\tau(p)$ mod ℓ for each of the six values of k listed above. Using methods which for the most part go back to Ramanujan, we proved all but one of these congruences -some of which were already known. For the remaining one, the notorious case of $\tau_{16}(p)$ mod 59, the numerical evidence is overwhelming; but although finite decision procedures are now known, no one has yet carried one of them through.

What was novel in [3] was the use of the Serre-Deligne theorem to prove that the potential congruences mod ℓ were restricted to certain types and that $\ell < 4k$ unless ℓ was a factor of the numerator of the Bernoulli number b_k. This left a finite list of hypothetical congruences to decide. The methods used to prove (or disprove) such congruences were essentially traditional, though put into a more polished and systematic form; and it may be noted that the one new type of congruence which we discovered is also the only one which we were unable to prove. In fact, the traditional methods of proving congruences mod ℓ are fairly straightforward; it is only the congruences modulo high powers of a prime which have in the past required the ingenious manipulation of complicated formulae. (See for example the list in [3], page 4 and the references there.)

Very little, and that unhelpful, was said in [3] about congruences modulo prime powers; but it was suggested that such congruences would be a by-product of any structure theory of modular forms mod ℓ^n that might be developed. Such a theory has since been obtained by Katz [1]; indeed in that paper he claims to give "an explicit solution" to the problem of finding all congruences mod ℓ which hold between the q-expansions of modular forms on $SL(2,\mathbf{Z})$. But this paper contains no examples, and it is not clear to me what he means by "explicit". On the other hand, I noticed during the Bonn conference that the Serre-Deligne theorem can be used not only to determine the possible forms of congruences for $\tau(p)$ mod ℓ^n but to prove those which are true; and the proofs obtained in this way are simpler and more systematic than the previous ones because they use much less information about the particular modular form f concerned. Indeed, all that is needed is the corresponding congruence for $\tau(p)$ mod ℓ, which is proved by the methods of [3], together with the values of $\tau(p)$ for certain particular p. For example, let

$$\Delta = q\{(1-q)(1-q^2)\dots\}^{24} = \sum \tau_{12}(n)q^n$$

be the normalized cusp form of weight 12; to prove Lahivi's congruence

$$\tau_{12}(p) \equiv p^{-30} + p^{41} \mod 5^3 \quad (p \neq 5) \qquad (2)$$

by these methods, all we need to know is that

$$\tau_{12}(p) \equiv p^2 + p \mod 5$$

and that (2) holds for $p = 2$ and $p = 11$.

The object of this paper is to explain the ideas involved and to apply them to obtain all congruences for $\tau(p) \mod \ell^n$ with $\ell \neq 2$ and $n > 1$, for the six values of k listed in the first paragraph. We assume throughout that $p \neq \ell$. Similar methods can probably be used for the case $\ell = 2$, but the details will be different and a glance at the Appendix to [3] makes it clear that they will be much more complicated. Ribet [2] has shown that the methods of [3] can be used even if the condition that the $\tau(n)$ are in $\mathbf{Z}$ is discarded; no doubt the same is true of the present paper.

For the reader's convenience we summarize the relevant parts of [3]. Let K_ℓ be the maximal algebraic extension of $\mathbf{Q}$ ramified only at ℓ. The Serre-Deligne theorem states that there is associated with each cusp form f satisfying the conditions in the first paragraph a continuous homomorphism

$$\rho_\ell : \mathrm{Gal}(K_\ell/\mathbf{Q}) \to GL_2(\mathbf{Z}_\ell)$$

such that ρ_ℓ (Frob(p)) has characteristic polynomial

$$X^2 - \tau(p)X + p^{k-1}.$$

If the image of ρ_ℓ is small enough, a knowledge of the determinant of an element of the image will imply some ℓ-adic information about the trace of that element; in particular, an approximate ℓ-adic knowledge of p will imply some ℓ-adic information about $\tau(p)$ and may therefore yield a congruence for $\tau(p) \mod \ell^n$. Conversely, the existence of such a congruence restricts the image of ρ_ℓ, because the set of

Frobenius elements is dense in the full Galois group. The image of $\det \circ \rho_\ell$ is $(\mathbf{Z}_\ell^*)^{k-1}$; so the image of ρ_ℓ is contained in the group of those elements of $GL_2(\mathbf{Z}_\ell)$ whose determinant is in $(\mathbf{Z}_\ell^*)^{k-1}$, and is equal to this group if and only if the image contains $SL_2(\mathbf{Z}_\ell)$. Consider the composite map

$$\tilde{\rho}_\ell : \mathrm{Gal}(K_\ell/\mathbf{Q}) \to GL_2(\mathbf{Z}_\ell) \to GL_2(\mathbf{F}_\ell)$$

where the right-hand arrow is reduction mod ℓ. For $\ell > 3$ it was shown in [3] that the image of ρ_ℓ contains $SL_2(\mathbf{Z}_\ell)$ if and only if the image of $\tilde{\rho}_\ell$ contains $SL_2(\mathbf{F}_\ell)$; and for $\ell = 2$ or 3 Tate has shown that neither of these can happen. We shall say that ℓ is an <u>exceptional prime</u> for f if the image of ρ_ℓ does not contain $SL_2(\mathbf{Z}_\ell)$, or equivalently if the image of $\tilde{\rho}_\ell$ does not contain $SL_2(\mathbf{F}_\ell)$.

There can only be a congruence for $\tau(p)$ mod ℓ if ℓ is an exceptional prime for f; this justifies replacing the search for congruence by the apparently more general search for exceptional primes. But conversely, from a knowledge of the image of $\tilde{\rho}_\ell$ it is easy to deduce what congruences $\tau(p)$ satisfies mod ℓ. So one of the main objects of [3] was to determine what are the possible images of $\tilde{\rho}_\ell$ and what congruences they correspond to. The conclusion can be stated as follows :

LEMMA 1. <u>Let ℓ be an exceptional prime for</u> f, <u>and denote by</u> G <u>the image of $\tilde{\rho}_\ell$; then</u> G, <u>and the associated congruence for</u> $\tau(p)$, <u>is of one of the following three kinds</u> :

(i) G <u>is contained in a Borel subgroup of</u> $GL_2(\mathbf{F}_\ell)$; <u>and there is an integer</u> m <u>such that</u>

$$\tau(p) \equiv p^m + p^{k-1-m} \mod \ell. \tag{3}$$

(ii) G <u>is contained in the normalizer of a Cartan subgroup of</u> $GL_2(\mathbf{F}_\ell)$ <u>but not in the Cartan subgroup itself</u>; <u>and</u> $\tau(p) \equiv 0 \mod \ell$ <u>whenever</u> p <u>is a quadratic non-residue</u> mod ℓ.

(iii) <u>The image of</u> G <u>in</u> $PGL_2(\mathbf{F}_\ell)$ <u>is isomorphic to the symmetric group</u> S_4; <u>and</u> $p^{1-k}\tau^2(p) \equiv 0,1,2$ <u>or</u> 4 mod ℓ.

<u>Case (ii) can only occur if</u> $\ell > 2$, <u>and case</u> (iii) <u>only if</u> $\ell > 3$.

In the rest of this paper we shall only be concerned with case (i). The example of case (ii) with $k = 12$, $\ell = 23$ is discussed in [3], pages 38-39, where in particular the exact image of ρ_ℓ and the best possible congruences for $\tau(p)$ are determined. That example is probably typical; certainly the other known example of this case (with $k = 16$, $\ell = 31$) behaves very similarly. As for case (iii), it seems best to postpone any further discussion until the congruences mod ℓ have been proved for the one suspected example of this case which has so far been found.

Most of the ideas in this paper were worked out at the Bonn conference. I am indebted to many of the participants, and particularly to Buhler, Coates, Serre, Tate and Wiles for useful conversations and for correcting my mistakes; and I am indebted to Atkin, Davenport and Stephens for providing numerical data without which the results below could not have been made explicit.

2. THE DIVISION INTO CASES.

Let χ_ℓ denote the canonical character

$$\chi_\ell : \mathrm{Gal}(K_\ell/\mathbf{Q}) \to \mathrm{Gal}(K_\ell^{ab}/\mathbf{Q}) \simeq \mathbf{Z}_\ell^*$$

which has the property that

$$\chi_\ell(\mathrm{Frob}(p)) = p,$$

and denote by $\tilde{\chi}$ the reduction of χ mod ℓ. The proof of (3) also shows that without loss of generality we may assume

$$\tilde{\rho}(\sigma) = \begin{pmatrix} \tilde{\chi}^m(\sigma) & * \\ 0 & \tilde{\chi}^{k-1-m}(\sigma) \end{pmatrix} \tag{4}$$

for all σ in $\mathrm{Gal}(K_\ell/\mathbf{Q})$; for details see [3], page 17. It is convenient to make two further normalizations, which will not affect the validity of (4).

For the first of these, we choose once for all a prime p_0 which is a generator of $\mathbf{Z}_\ell^*$ regarded as a topological group. If the equation

$$\tau(p_0) = x + p_0^{k-1}x^{-1} \tag{5}$$

is regarded momentarily as a congruence mod ℓ, its two roots are known to be p_0^m and p_0^{k-1-m}; these are incongruent because $(\ell-1)$, being even, does not divide $(k-1-2m)$. Hence (5) has a root x_0 in $\mathbf{Z}_\ell$ such that

$$x_0 \equiv p_0^m \quad \mathrm{mod}\ \ell, \tag{6}$$

by Hensel's Lemma; and the eigenvalues of $\rho_\ell(\mathrm{Frob}(p_0))$ are x_0 and $p_0^{k-1}x_0^{-1}$. There is therefore a matrix M_0 in $GL_2(\mathbf{Z}_\ell)$ such that

$$M_0^{-1}\rho_\ell(\mathrm{Frob}(p_0))M_0 = \begin{pmatrix} x_0 & 0 \\ 0 & p_0^{k-1}x_0^{-1} \end{pmatrix};$$

and it follows from (4) and (6) that $\widetilde{M}_0$ is upper triangular. Replacing the representation ρ_ℓ by $M_0^{-1}\rho_\ell M_0$, which does not affect (4), we may assume

$$\rho_\ell(\mathrm{Frob}(p_0)) = \begin{pmatrix} x_0 & 0 \\ 0 & p_0^{k-1}x_0^{-1} \end{pmatrix}. \tag{7}$$

For any z in $\mathbf{Z}_\ell^*$ and any μ in

$$R = \varprojlim \mathbf{Z}/(\ell^n(\ell-1))$$

there is a natural definition of z^μ as an element of $\mathbf{Z}_\ell^*$, and this has all the properties which the notation suggests. It follows from (6) that there exists μ in R such that

$$x_0 = p_0^{\mu} \quad \text{and} \quad \mu \equiv m \quad \mod(\ell-1) \tag{8}$$

and we can clearly replace m by μ in (3) and (4) without affecting their validity. But any z in $\mathbf{Z}_\ell^*$ can be expressed as $\lim p_0^n$ where n runs through a suitably chosen sequence of elements of $\mathbf{Z}$; and by (7) and (8) the corresponding sequence of matrices $\rho_\ell((\mathrm{Frob}(p_0))^n)$ tends to

$$\begin{pmatrix} z^{\mu} & 0 \\ 0 & z^{k-1-\mu} \end{pmatrix}. \tag{9}$$

Since $\mathrm{Gal}(K_\ell/\mathbf{Q})$ is compact, the image of ρ_ℓ is closed; so the matrix (9) is in the image of ρ_ℓ for every z in $\mathbf{Z}_\ell^*$. Moreover, by the way it has been constructed, (9) must be equal to $\rho_\ell(\sigma)$ for some σ with $\chi(\sigma) = z$. The importance of this is that it allows us to work with

$$\Gamma = \{\rho_\ell(\sigma) \mid \chi(\sigma) = 1\}$$

instead of with the full image of ρ, since every coset of Γ in the image of ρ contains a matrix of the form (9).

Before the second normalization, we must clear out of the way one highly implausible special case.

LEMMA 2. *The image of ρ_ℓ is not contained in any Borel subgroup of $GL_2(\mathbf{Z}_\ell)$.*

PROOF. Suppose the lemma is false; then we may without loss of generality assume that the image of ρ_ℓ consists of upper triangular matrices, say

$$\rho_\ell(\sigma) = \begin{pmatrix} a(\sigma) & b(\sigma) \\ 0 & d(\sigma) \end{pmatrix}.$$

Now $\sigma \mapsto a(\sigma)$ and $\sigma \mapsto d(\sigma)$ are continuous homomorphisms $\mathrm{Gal}(K_\ell/\mathbf{Q}) \to \mathbf{Z}_\ell^*$

and hence must be powers of χ_ℓ, by class-field theory. Hence for some μ, which is in fact the same as in (9), we have

$$\tau(p) = p^\mu + p^{k-1-\mu}; \tag{10}$$

and by restricting ourselves to primes $p \equiv 1 \mod \ell$ we can assume that μ is in $\mathbf{Z}_\ell$ rather than in R. But (10) implies that p^μ is algebraic over $\mathbf{Q}$; hence by a theorem of Lang μ is rational, and since $\tau(p)$ is in $\mathbf{Z}$, it follows from (10) that μ is an integer with $0 \leqslant \mu \leqslant k-1$. Now (1) gives

$$f = \sum n^\mu \sigma_{k-1-2\mu}(n) q^n$$

which is in general a derivative of an Eisenstein series and certainly is never a cusp form. This contradiction proves the lemma.

We may therefore assume that if we write

$$\rho_\ell(\sigma) = \begin{pmatrix} a(\sigma) & b(\sigma) \\ c(\sigma) & d(\sigma) \end{pmatrix},$$

as we shall consistently do from now on, then neither $b(\sigma)$ nor $c(\sigma)$ vanishes identically. Let $n \geqslant 0$ be the largest integer such that ℓ^n divides $b(\sigma)$ for every σ, and write $\Lambda = \mathrm{diag}(\ell^n,1)$; then we can replace the representation ρ_ℓ by $\Lambda^{-1}\rho_\ell\Lambda$ without affecting what we have done so far. This is equivalent to multiplying $c(\sigma)$ and dividing $b(\sigma)$ by ℓ^n; so it ensures that not all $b(\sigma)$ are divisible by ℓ. Clearly this remains true even if we restrict ourselves to Γ. This completes the normalization of ρ.

Let $N_1 > 0$ and $N_2 > 0$ be the largest integers such that

$$a(\sigma) \equiv d(\sigma) \equiv 1 \mod \ell^{N_1} \text{ and } c(\sigma) \equiv 0 \mod \ell^{N_2} \tag{11}$$

for all $\rho(\sigma)$ in Γ. For the moment N_1 may be infinite; but N_2 is finite by Lemma 2. Since anything in the image of ρ can be written as a product of an element of Γ with a matrix of the form (9), the second congruence (11) holds for all σ in $\mathrm{Gal}(K_\ell/\mathbf{Q})$.

LEMMA 3. $N_1 = N_2$ and both are finite.

PROOF. Consider the identity

$$\begin{pmatrix} a_1 & b_1 \\ c_1 & d_1 \end{pmatrix}\begin{pmatrix} a_2 & b_2 \\ c_2 & d_2 \end{pmatrix} = \begin{pmatrix} a_1a_2+b_1c_2 & * \\ * & * \end{pmatrix}$$

in which the two matrices on the left are in Γ and are chosen that $\ell \nmid b_1$ and $\ell^{N_2+1} \nmid c_2$. Clearly a_1, a_2 and $a_1a_2+b_1c_2$ cannot all be congruent to $1 \mod \ell^{N_2+1}$; so $N_1 \leq N_2$. Suppose now that $N_1 < N_2$; then the continuous map

$$\sigma \to a(\sigma) \mod \ell^{N_1+1}$$

induces a homomorphism $\mathrm{Gal}(K_\ell/\mathbf{Q}) \to (\mathbf{Z}/\ell^{N_1+1})^*$. Since the image is commutative, this map must factor through χ_ℓ and hence all σ with $\chi(\sigma) = 1$ lie in its kernel. This means that $a(\sigma) \equiv 1 \mod \ell^{N_1+1}$ for all $\rho(\sigma)$ in Γ, contrary to the definition of N_1; and this contradiction completes the proof of the lemma. Henceforth we shall write

$$N_1 = N_2 = N \geq 1;$$

it is not hard to see that although our normalization depends on the choice of p_0, the value of N depends only on f and ℓ.

Let $\tilde{G}$ denote the group of elements $\{\alpha,\beta,\gamma,\delta\}$ with $\alpha,\beta,\gamma,\delta$ in $\mathbf{F}_\ell$ and subject to the condition

$$\alpha + \delta = \beta\gamma; \tag{12}$$

the group law in $\tilde{G}$ is to be given by

$$\{\alpha_1,\beta_1,\gamma_1,\delta_1\}\{\alpha_2,\beta_2,\gamma_2,\delta_2\} = \{\alpha_1+\alpha_2+\beta_1\gamma_2, \beta_1+\beta_2, \gamma_1+\gamma_2, \delta_1+\delta_2+\gamma_1\beta_2\}.$$

It is easy to check that the map

$$\rho(\sigma) \to \{\ell^{-N}(a-1), b, \ell^{-N}c, \ell^{-N}(d-1)\},$$

where the components on the right are to be replaced by their images in $\mathbf{F}_\ell$, defines a homomorphism $\Gamma \to \tilde{G}$. Indeed this motivates the definition of $\tilde{G}$; in particular the condition (12) corresponds to ad-bc = 1. Let $\tilde{\Gamma}$ be the image of Γ in $\tilde{G}$; our next task is to list the possibilies for $\tilde{\Gamma}$. There is an exact sequence

$$0 \to C_\ell \to \tilde{G} \to C_\ell \times C_\ell \to 0,$$

where C_ℓ denotes the cyclic group of order ℓ and the two inner maps are defined by

$$\alpha \to \{\alpha, 0, 0, -\alpha\} \quad \text{and} \quad \{\alpha\ \beta\ \gamma\ \delta\} \to \beta \times \gamma;$$

moreover $\tilde{G}$ is not commutative. By hypothesis, neither β nor γ can vanish identically on $\tilde{\Gamma}$; straightforward calculation therefore shows that $\tilde{\Gamma}$ must be one of the following three types :

(a) the whole group $\tilde{G}$;

(b) the elements with $\beta = \lambda\gamma$, for some fixed λ in F_ℓ^*;

(c) the cyclic group generated by $\{\alpha_0, \beta_0, \gamma_0, \delta_0\}$, where $\beta_0\gamma_0 \neq 0$.

In these three cases the order of $\tilde{\Gamma}$ will be ℓ^3, ℓ^2 or ℓ respectively. In connection with (c) we note that

$$\{\alpha,\beta,\gamma,\delta\}^n = \{n\alpha+\tfrac{1}{2}n(n-1)\beta\gamma, n\beta, n\gamma, n\delta+\tfrac{1}{2}n(n-1)\beta\gamma\} \tag{13}$$

as can easily be verified by induction.

<u>THEOREM</u> 1. <u>The group</u> $\tilde{\Gamma}$ <u>is of type (c) if</u>

$$2(k-1-2m) \equiv 0 \mod(\ell-1) \tag{14}$$

<u>and of type (a) otherwise</u>. <u>It is never of type (b)</u>.

PROOF. By (8), we can write μ for m in (14). Any σ in $\mathrm{Gal}(K_\ell/\mathbf{Q})$ acts on Γ by conjugation by $\rho(\sigma)$, and this induces an automorphism of $\tilde{\Gamma}$. If $\rho(\sigma)$ is given by (9), then this automorphism is

$$\{\alpha,\beta,\gamma,\delta\} \to \{\alpha,\beta\nu,\gamma\nu^{-1},\delta\} \tag{15}$$

where ν is the image of $z^{k-1-2\mu}$ in $\mathbf{F}_\ell^*$. If (14) is false we can choose z so that $\nu \neq \pm 1$; and in cases (b) and (c) this means that (15) does not map $\tilde{\Gamma}$ to itself. So if (14) is false $\tilde{\Gamma}$ must be of type (a).

Suppose therefore that (14) is true, and write $K = \mathbf{Q}(\sqrt{-\ell})$. Since k is even, $\ell \equiv -1 \mod 4$; so K is the unique quadratic subfield of K_ℓ. Suppose σ is in

$$\mathrm{Gal}(K_\ell/K) = \{\sigma \text{ in } \mathrm{Gal}(K_\ell/\mathbf{Q}) \mid \chi^{(\ell-1)/2}(\sigma) \equiv 1 \mod \ell\};$$

then $\nu = 1$ and the automorphism (15) is the identity. Hence there is a homomorphism

$$\mathrm{Gal}(K_\ell/K) \to \tilde{\Gamma}/[\tilde{\Gamma},\tilde{\Gamma}] \times E, \tag{16}$$

where the square brackets denote the commutator subgroup and $E \simeq \mathbf{Z}$ is the group of $z \equiv 1 \mod \ell$ in $\mathbf{Z}_\ell^*$; this homomorphism is given by

$$\sigma \mapsto (\text{Image of } \rho(\sigma)\mathrm{Diag}(\chi^{-\mu}(\sigma),\chi^{\mu+1-k}(\sigma))) \times \chi^{\ell-1}(\sigma)$$

and it is easily seen to be onto. In cases (a) and (b),

$$\tilde{\Gamma}/[\tilde{\Gamma},\tilde{\Gamma}] \simeq C_\ell \times C_\ell. \tag{17}$$

But crude estimates show that the class number of K is less than ℓ and prime to ℓ; so class-field theory tells us that any commutative ℓ-group which is a quotient of $\mathrm{Gal}(K_\ell/K)$ has rank at most 2. Since (16) is onto, (17) is impossible; and therefore $\tilde{\Gamma}$ must be of type (c).

COROLLARY. *If $\tilde{\Gamma}$ is of type (c) then $\alpha_0 = \delta_0 = \frac{1}{2}\beta_0\gamma_0$; in other words* Γ

consists of the $\{\frac{1}{2}\lambda n^2, \lambda n, n, \frac{1}{2}\lambda n^2\}$ where λ in $\mathbf{F}_\ell^*$ is fixed and n runs through the elements of $\mathbf{F}_\ell$.

PROOF. It follows from (14) that

$$k-1-2\mu \equiv \tfrac{1}{2}(\ell-1) \mod(\ell-1), \tag{18}$$

since the left hand side is odd; so in the notation of (15) we can choose z so that $\nu = -1$. Comparison with (13) now gives

$$\{\alpha_0, -\beta_0, -\gamma_0, \delta_0\} = \{\alpha_0, \beta_0, \gamma_0, \delta_0\}^{-1}$$

and therefore $\alpha_0 = \delta_0 = \frac{1}{2}\beta_0\gamma_0$. This proves the Corollary.

In any particular example it is easy to check whether (14) holds and hence to find whether $\tilde{\Gamma}$ is of type (a) or type (c). At first sight type (c) breaks up into $(\ell-1)$ subtypes, according to the value of λ in the Corollary; but we are still allowed to replace ρ by $R^{-1}\rho R$ where $R = \mathrm{Diag}(r,1)$ with r in $\mathbf{Z}_\ell^*$, and this operation replaces λ by $r^2\lambda$. So there are really only two subtypes, according as λ is a square or not.

At this point the argument divides, according as we are in case (a) or case (c). For each case we wish to find the associated congruences for $\tau(p)$, which will of course involve μ and N; and we need to prove a constructive method of finding the value of N for any particular f. It would be desirable also to find the exact image of ρ, but I have been able to achieve this only in case (a). The 42 known examples of case (i) of Lemma 1 are listed in [3], page 32. Of these, 6 have $\ell = 2$ and hence fall outside the scope of this paper, 25 do not satisfy (14) and are therefore of type (a), and 11 do satisfy (14) and are therefore of type (c). Details of these 36 examples are given below.

3. FURTHER INVESTIGATION OF CASE (a).

In case (a) we start by determining Γ, which is equivalent to determining the image of ρ. It turns out that Γ is as large as possible, subject to (11) :

THEOREM 2. In case (a),

$$\Gamma = \{\begin{pmatrix} a & b \\ c & d \end{pmatrix} \text{ in } SL_2(\mathbf{Z}_\ell) \mid a-1 \equiv c \equiv d-1 \equiv 0 \mod \ell^N\}.$$

PROOF. For $r = 0,1,2,\ldots$ let G_r be the set of matrices $\begin{pmatrix} a & b \\ c & d \end{pmatrix}$ in $SL_2(\mathbf{Z}_\ell)$ which satisfy

$$a-1 \equiv c \equiv d-1 \equiv 0 \mod \ell^{N+r}, \ b \equiv 0 \mod \ell^r$$

and let S_r be the set of matrices $\begin{pmatrix} \alpha & \beta \\ \gamma & \delta \end{pmatrix}$ in $M_2(\mathbf{Z}_\ell)$ which satisfy

$$\alpha \equiv \gamma \equiv \delta \equiv 0 \mod \ell^{N+r}, \ \beta \equiv 0 \mod \ell^r,$$

thus g in G_r implies $(g-I)$ in S_r. It is easy to see that G_r is a group, that G_{r+1} is a normal subgroup of G_r of index ℓ^3, and that the G_r form a base for the neighbourhoods of I in $SL_2(\mathbf{Z}_\ell)$. Moreover Γ is closed. So to prove the theorem it is enough to prove that Γ contains representatives of each coset of G_{r+1} in G_r for $r = 0,1,2,\ldots$. This is true for $r = 0$, for the natural map $\Gamma \to \tilde{G} \simeq G_0/G_1$ is onto since we are in case (a). We therefore proceed by induction, assuming the result for $r-1$ and proving it for r; thus we can take $r \geqslant 1$.

Write $g = \begin{pmatrix} a & b \\ c & d \end{pmatrix}$ and let g be in G_{r-1}. If we further assume that $\ell | b$, it follows easily that

$$(g-I)^n \text{ is in } S_{r+n-2} \text{ for } n = 1,2,\ldots ;$$

and since $\ell > 2$ this implies

$$g^\ell = I + \ell(g-I) + \tfrac{1}{2}\ell(\ell-1)(g-I)^2 + \ldots + (g-I)^\ell$$
$$\equiv I + \ell(g-I) \mod S_{r+1}.$$

Thus if g_1 and g_2 are in distinct cosets of G_r in G_{r-1}, then g_1^ℓ and g_2^ℓ are in distinct cosets of G_{r+1} in G_r. For $r > 1$ this completes the induction step, because the condition $\ell|b$ is then automatic.

For $r = 1$ this argument only shows that Γ meets those ℓ^2 cosets of G_2 in G_1 for which $\ell^2|b$. However, a similar argument shows that it also meets those cosets for which $\ell^{N+2}|c$; and so Γ meets more than ℓ^2 cosets of G_2 in G_1. The cosets which Γ meets form a subgroup of G_1/G_2, and since this has order ℓ^3 it has no proper subgroup of order greater than ℓ^2. This completes the induction step for $r = 1$, and thus also completes the proof of the theorem.

COROLLARY 1. *The image of* ρ_ℓ *consists of all* $\begin{pmatrix} a & b \\ c & d \end{pmatrix}$ *in* $GL_2(\mathbf{Z}_\ell)$ *such that*

$$a-z^\mu \equiv c \equiv d-z^{k-1-\mu} \equiv 0 \mod \ell^N$$

for some z *in* $\mathbf{Z}_\ell^*$.

PROOF. The matrix (9) is in the image of ρ for every such z.

COROLLARY 2. *The coefficients* $\tau(p)$ *satisfy, for* $p \neq \ell$,

$$\tau(p) \equiv p^\mu + p^{k-1-\mu} \mod \ell^N \tag{19}$$

and in this congruence μ *can be replaced by any integer* $m \equiv \mu \mod \ell^{N-1}(\ell-1)$. *Moreover this result is best possible in the sense that given any* τ *in* $\mathbf{Z}_\ell$ *and* π *in* $\mathbf{Z}_\ell^*$ *with* $\tau \equiv \pi^\mu + \pi^{k-1-\mu} \mod \ell^N$, *we can find a prime* p *such that* $\tau(p)-\tau$ *and* $p-\pi$ *are both as small as we like in the* ℓ*-adic metric*.

PROOF. The congruence (19) is just the relation between trace and determinant which holds for every element of the image of ρ, by Corollary 1; and the rest of the first sentence is trivial. The second sentence follows from Corollary 1 and the fact that Frobenius elements are dense in $\mathrm{Gal}(K_\ell/\mathbf{Q})$.

It remains only to give a method of calculating N for any given f; for the definition of N above was totally non-constructive. To prove $N = 1$ it is enough to find a prime p such that

$$\tau(p) \not\equiv p^{\mu} + p^{k-1-\mu} \mod \ell^2,$$

and here we may replace μ by any $m \equiv \mu \mod \ell(\ell-1)$. This shortcuts the more elaborate argument below; and it enables one to show that of the 25 known examples of case (a), the 17 with $\ell \geqslant 11$ all have $N = 1$.

The systematic calculation of N depends on finding a finite list of primes p_r (the list depending only on ℓ) for at least one of which

$$\tau(p_r) \not\equiv p_r^{\mu} + p_r^{k-1-\mu} \mod \ell^{N+1}. \tag{20}$$

Write $L = \mathbf{Q}(\sqrt[\ell]{1})$ and assume that ℓ is a regular prime - that is, the class number of L is prime to ℓ; it was to permit this assumption that the ad hoc test for $N = 1$ was given above. Since

$$\mathrm{Gal}(K_\ell/L) = \{\sigma \text{ in } \mathrm{Gal}(K_\ell/\mathbf{Q}) \mid \chi(\sigma) \equiv 1 \mod \ell\},$$

there is a homomorphism

$$\mathrm{Gal}(K_\ell/L) \to C_\ell \times C_\ell \times C_\ell \tag{21}$$

which is given by

$$\sigma \mapsto (b(\sigma) \mod \ell) \times (\ell^{-N}c(\sigma) \mod \ell) \times (\chi(\sigma) \mod \ell^2); \tag{22}$$

and Theorem 2 shows that this is onto. Let μ_ℓ denote the group of ℓth roots of unity, and $\hat{L}$ the maximum abelian ℓ-extension of L which is unramified outside ℓ and for which $Gal(\hat{L}/L)$ is annihilated by ℓ. The map (21) factors through $Gal(\hat{L}/L)$; and Kummer theory states that

$$U/U^{\ell} \simeq Hom(Gal(\hat{L}/L),\mu_{\ell}) \tag{23}$$

is an isomorphism of $Gal(L/\mathbf{Q})$ modules, where U is the multiplicative group of elements of L which are units outside of ℓ. Clearly (23) implies

$$Gal(\hat{L}/L) \simeq C_{\ell}^{(\ell+1)/2} \tag{24}$$

as abelian groups; and it is known that the map (24) can be written as

$$\sigma \mapsto \xi_0(\sigma) \times \xi_1(\sigma) \times \ldots \times \xi_{\ell-2}(\sigma)$$

where the non-zero subscripts are the odd integers. In this notation α in $Gal(L/\mathbf{Q})$ acts on $\xi_{\nu}(\sigma)$ according to the rule

$$\xi_{\nu}(\alpha\sigma\alpha^{-1}) = \alpha^{\nu}\xi_{\nu}(\sigma), \tag{25}$$

where on the right α has been identified with its image in $\mathbf{F}_{\ell}^{*}$. In particular, ξ_0 corresponds to χ; to be precise

$$\xi_0(\sigma) \equiv \ell^{-1}(\chi(\sigma)-1) \mod \ell,$$

so that the third factor in (21) is just ξ_0.

Take $p \equiv 1 \mod \ell$ and write $\rho(\mathrm{Frob}(p)) = \begin{pmatrix} a & b \\ c & d \end{pmatrix}$; then

$$\begin{aligned} \tau(p) &= a+d = 1+p^{k-1}+bc-(a-1)(d-1) \\ &\equiv 1+p^{k-1}+bc-(p^{\mu}-1)(p^{k-1-\mu}-1) \mod \ell^{N+1} \\ &\equiv p^{\mu}+p^{k-1-\mu}+bc \mod \ell^{N+1}. \end{aligned}$$

Denote the first two factors on the right of (22) by $x_1(\sigma)$ and $x_2(\sigma)$ respectively. This last calculation shows that

$$x_1(\sigma)x_2(\sigma) \text{ is fixed under the action of } Gal(L/\mathbf{Q}) \tag{26}$$

at least when σ = Frob(p) - and hence in general since the action is continuous and the Frob(p) are dense in Gal(K_ℓ/L). The reader who is confused by the implied reference to the action of Gal(L/**Q**) on Frob(p) may be helped by the following explanation. Though by abuse of language one speaks of Frob(p) as an element of Gal(K_ℓ/**Q**), it is in fact an entire conjugacy class of such elements. If $p \equiv 1 \mod \ell$, all these elements lie in Gal(K_ℓ/L); but there they form $(\ell-1)$ conjugacy classes, corresponding to the $(\ell-1)$ prime factors of p in L. By the same abuse of language, these are described as the $(\ell-1)$ Frobenii of p for Gal(K_ℓ/L); and Gal(L/**Q**) permutes them in the same way as it permutes the prime factors of p in L.

Now x_1 and x_2 are linear combinations of the ξ_ν; and x_1, x_2 and ξ_0 are linearly independent because (21) is onto. Let α_0 be a generator of Gal(L/**Q**); by (26), α_0 either takes each of x_1 and x_2 into a multiple of itself or into a multiple of the other. In either case α_0^2 takes each of x_1 and x_2 into a multiple of itself. Suppose first that $\ell \equiv -1 \mod 4$; then the only case where two of the $\alpha_0^{2\nu}$ arising from (25) can be equal is when $\nu = 0$ or $\frac{1}{2}(\ell-1)$. Hence either each of x_1 and x_2 is a multiple of some ξ_ν or

$$x_1 = u\xi_0 + v\xi_{(\ell-1)/2}, \quad x_2 = u\xi_0 - v\xi_{(\ell-1)/2}$$

for some non-zero u,v. The second alternative is impossible, by the linear independence of x_1, x_2 and ξ_0; so (26) implies

$$x_1 x_2 = u\xi_{2r-1}\xi_{\ell-2r} \tag{27}$$

for some non-zero u and some r with $0 < r < (\ell+1)/4$. We cannot have $r = (\ell+1)/4$ because that would again contradict linear independence. Now suppose instead that $\ell \equiv 1 \mod 4$. The same argument about the action of α_0^2 shows that either each of x_1 and x_2 is a multiple of some ξ_ν or

$$x_1 = u\xi_r + v\xi_s,\ x_2 = u\xi_r - v\xi_s \text{ with } s = r + \tfrac{1}{2}(\ell-1)$$

for non-zero u,v and some odd r. The second alternative is again impossible, this time because it contradicts (26); so again we have (27).

What we have just proved can be summarized as follows :

THEOREM 3. *Suppose that we are in case (a) and let the ξ_ν be as above. For each r with $0 < r < (\ell+1)/4$ let $p_r \equiv 1 \mod \ell$ be such that neither $\xi_{2r-1}(\sigma)$ nor $\xi_{\ell-2r}(\sigma)$ vanishes for $\sigma = \mathrm{Frob}(p_r)$. Then (20) holds for at least one of the p_r.*

For given f and ℓ, (19) and (20) provide a completely constructive method of calculating N once we have obtained a suitable list of p_r. It is also possible to calculate the value of r in (27), though I have not in fact done so; but it seems unlikely that anything of interest would emerge. To obtain a list of suitable p_r we proceed as follows. To each ξ_ν there corresponds a field $L_\nu \subset \hat{L}$, of degree ℓ over L, such that $\xi_\nu(\mathrm{Frob}(p)) = 0$ if and only if p splits completely in L_ν. If $L_\nu = L(\sqrt[\ell]{u_\nu})$ with u_ν in U, then (23) implies that

$$(\alpha u_\nu)/u_\nu^n \text{ is an } \ell^{th} \text{ power in L,} \tag{28}$$

where α is any element of Gal(L/**Q**) and n is the image of $\alpha^{1-\nu}$ in $\mathbf{F}_\ell^*$. Conversely, if u_ν satisfies (28) for every α, this gives a recipe for finding L_ν. In particular we can take $u_0 = \sqrt[\ell]{1}$ and $u_1 = \ell$.

The two cases that concern us here are $\ell = 5$ and $\ell = 7$, for each of which r = 1 is forced. We can choose

$$u_3 = \tfrac{1}{2}(1+\sqrt{5}) \quad \text{for } \ell = 5$$

$$u_3 = \varepsilon_1\varepsilon_2^4\varepsilon_3^2 \text{ and } u_5 = \varepsilon_1\varepsilon_2^2\varepsilon_3^4 \quad \text{for } \ell = 7$$

where $\varepsilon_\nu = 2\cos 2\pi i\nu/7$; it is easy to check that these satisfy (28). Moreover $p = 11$ does not split in L_1 or L_3 for $\ell = 5$; and $p = 29$ does not split in L_1 or L_5 for $\ell = 7$. Hence we have

COROLLARY. *Condition (20) holds for* $\ell = 5$, $p = 11$ *and for* $\ell = 7$, $p = 29$.

Explicit calculation now gives the following results :

THEOREM 4. *We have* $N > 1$ *for the following 8 examples of case (a),*

ℓ	5	5	5	5	5	5	7	7
k	12	16	18	20	22	26	16	22
N	3	2	3	2	2	2	3	2
μ	41	17	22	13	14	6	85	37

where μ *is to be taken* mod $\ell^{N-1}(\ell-1)$.

4. FURTHER INVESTIGATION OF CASE (c).

In case (a) we were able to determine Γ quite easily, because there was only one way of lifting $\tilde{\Gamma}$ back to a closed subset of $SL_2(\mathbf{Z}_\ell)$. In case (c) this is no longer true; indeed even the extra information that the matrices (9) lie in the image of ρ is not enough to determine Γ uniquely. The best substitute for Theorem 2 and its first Corollary that we have is that for fixed λ and some n

$$\rho(\sigma) \equiv \begin{pmatrix} (1+\frac{1}{2}\lambda n^2\ell^N)\chi^\mu & \chi^\mu\lambda n \\ \chi^{k-1-\mu}n\ell^N & (1+\frac{1}{2}\lambda n^2\ell^N)\chi^{k-1-\mu} \end{pmatrix} \tag{29}$$

where the congruences for a,c,d are to be taken mod ℓ^{N+1} and that for b is to be taken mod ℓ; moreover for given χ and n we can find σ such that these congruences hold. All this is just a restatement of the Corollary to Theorem 1. It follows from (29) that

$$\tau(p) \equiv (p^{\mu} + p^{k-1-\mu})(1 + \tfrac{1}{2}\lambda n^2 \ell^N) \quad \mathrm{mod}\ \ell^{N+1}; \tag{30}$$

and remembering (18) this gives the following result.

THEOREM 5. *Assume we are in case (c), with* $p \neq \ell$; *then*

$$\tau(p) \equiv p^{\mu} + p^{k-1-\mu} \quad \begin{array}{ll} \mathrm{mod}\ \ell^N & \textit{if}\ (\frac{p}{\ell}) = +1, \\ \mathrm{mod}\ \ell^{N+1} & \textit{if}\ (\frac{p}{\ell}) = -1. \end{array} \tag{31}$$

In this congruence μ *can be replaced by any integer* $m \equiv \mu \mod \ell^N(\ell-1)$.

We have nothing here analogous to the last sentence of Corollary 2 to Theorem 2. For fixed $p \mod \ell^N$ with $(\frac{p}{\ell}) = +1$, exactly $\frac{1}{2}(\ell+1)$ of the ℓ residue classes mod ℓ^{N+1} allowed by (31) actually occur; when $(\frac{p}{\ell}) = -1$ we can say nothing. However, this theorem does fit the pattern of the known result for τ_{12} when $\ell = 3$ or 7; see for example [3], page 4.

As in §3 we can give an ad hoc method of proving $N = 1$ which short-cuts the more elaborate arguments below; this is less necessary than in case (a) because $\ell < 2k$ in case (c), but it still saves effort. Suppose that $(\frac{p}{\ell}) = +1$; then to alter μ by a multiple of $(\ell-1)$ does not alter $p^{\mu} + p^{k-1-\mu} \mod \ell^2$. So to prove $N = 1$ it is enough to find p,m with

$$\tau(p) \not\equiv p^m + p^{k-1-m} \mod \ell^2, \ m \equiv \mu \mod (\ell-1), (\tfrac{p}{\ell}) = +1.$$

This test is enough to prove that $N = 1$ for the cases

$$\ell = 7,\ k = 12,18,20,26 \text{ and } \ell = 11,\ k = 18,$$

leaving for systematic calculation only the 6 cases with $\ell = 3$.

For the systematic calculation of N in this case we consider

$$\mathrm{Gal}(K_\ell/K) = \{\sigma \text{ in } \mathrm{Gal}(K_\ell/\mathbf{Q}) \mid \chi^{k-1-2\mu}(\sigma) \equiv 1 \mod \ell\}$$

where $K = \mathbf{Q}(\sqrt{-\ell})$ as in §2. There is a homomorphism

$$\mathrm{Gal}(K_\ell/K) \to C_\ell \times C_\ell \tag{32}$$

which is given by

$$\sigma \mapsto (\chi^{-\mu}(\sigma)b(\sigma) \mod \ell) \times (\chi^{\ell-1}(\sigma) \mod \ell^2); \tag{33}$$

and the remarks following (29) show that this is onto. Let $x(\sigma)$ denote the first factor on the right of (33); then (29) and (30) imply

$$\tau(p) \equiv (p^\mu + p^{k-1-\mu})(1+\tfrac{1}{2}\lambda^{-1}x^2\ell^N) \mod \ell^{N+1} \tag{34}$$

where $x = x(\mathrm{Frob}(p))$. This equation determines x up to sign, so just as in §3 the non-trivial automorphism of K over $\mathbf{Q}$ must either fix $x(\sigma)$ or take it to $-x(\sigma)$. In the former case $x(\sigma)$ would be a function of $\chi(\sigma)$, which is impossible because (32) is onto; so the latter case must hold. This is enough to determine $x(\sigma)$ up to multiplication by a non-zero constant. Thus if we choose p so that $x \neq 0$, then (34) provides a completely constructive way of finding N.

In particular, when $\ell = 3$ and $p \equiv 1 \mod 3$ we can write

$$4p = u^2 + 27v^2 = 4\pi\bar{\pi} \tag{35}$$

in essentially only one way, where $2\pi = u+3v\sqrt{-3}$, and then

$$x(\mathrm{Frob}(\pi)) = \pm\, v/u \mod 3$$

for a fixed choice of sign. So for $\ell = 3$ we can take $p = 7$ in (20); and indeed (34) gives

$$\tau(p) \equiv (p^\mu + p^{k-1-\mu})(1 \pm 3^N v^2) \mod 3^{N+1} \tag{36}$$

where the sign depends only on f. This gives the following table.

k	12	16	18	20	22	26
N	6	5	5	5	6	5
μ mod 2.3^N	848	174	386	298	18	340
sign in (36)	+	+	-	-	-	-

There are similar recipes for the cases $\ell = 7$ and $\ell = 11$.

However, once we are willing to take into our congruences the factors of p in K (or the expression of p by the appropriate quadratic form, which is equivalent), we can do better than (34). The first evidence for this was the conjecture, based on extensive numerical evidence, that in the notation of (35)

$$\tau_{12}(p) \equiv p^{119} + p^{-108} - 3^6 v^2 \mod 3^8 \qquad (37)$$

and not merely mod 3^7 as (34) and the table following (36) would imply. This has been discovered more than once; see for example [3], page 42. On similar grounds Atkin has conjectured that

$$\tau_{12}(p) \equiv p + p^{10} - 21v^2 \mod 7^2 \qquad (38)$$

whenever

$$4p = u^2 + 7v^2; \qquad (39)$$

both these conjectures will be proved below. There are obvious historical reasons why τ_{12} has been studied more intensively than the other τ_k, but there is no other reason why there should not have been similar conjectures for other values of k. Results of this kind can be fitted into the framework of the present paper, though the process is somewhat untidy - reflecting perhaps the diversity of possible images of ρ_ℓ. The 3-adic results, for all which $N > 1$, depend on the following fact :

LEMMA 4. *Suppose that we are in case (c), and that $N > 1$; and let λ denote any pull-back to $\mathbb{Z}$ of the λ in (29). Then the map*

$$\sigma \mapsto b(\sigma)\chi^{-\mu}(\sigma) + \lambda\ell^{-N}c(\sigma)\chi^{\mu+1-k}(\sigma) \tag{40}$$

induces a homomorphism

$$\mathrm{Gal}(K_\ell/K) \to \mathbf{Z}/(\ell^2). \tag{41}$$

PROOF. Denote the map (40) by f, and write $a(\sigma_i) = a_i$, $\chi(\sigma_i) = \chi_i$ and so on for $i = 1$ or 2. We have

$$f(\sigma_1\sigma_2) - f(\sigma_1) - f(\sigma_2) = b_2\chi_2^{-\mu}(a_1\chi_1^{-\mu}-1) + \lambda\ell^{-N}c_2\chi_2^{\mu+1-k}(d_1\chi_1^{\mu+1-k}-1)$$
$$+ b_1\chi_1^{-\mu}(d_2\chi_2^{-\mu}-1) + \lambda\ell^{-N}c_1\chi_1^{\mu+1-k}(a_2\chi_2^{\mu+1-k}-1).$$

The first two terms on the right are divisible by ℓ^N, by (29); and for the same reason we have

$$d_2\chi_2^{-\mu} \equiv \nu_2,\ a_2\chi_2^{\mu+1-k} \equiv \nu_2^{-1} \mod \ell^N \tag{42}$$

where $\nu_2 = \chi_2^{k-1-2\mu}$. Hence the right hand side is congruent mod ℓ^N to

$$(\nu_2-1)(b_1\chi_1^{-\mu}-\nu_2^{-1}\lambda\ell^{-N}c_1\chi_1^{\mu+1-k}).$$

Each of these two factors is divisible by ℓ, by (29) again, and so

$$f(\sigma_1\sigma_2) - f(\sigma_1) - f(\sigma_2) \equiv 0 \mod \ell^2$$

since $N > 1$. This completes the proof of the lemma.

To identify the homomorphism (41) we recall that there are two independent abelian characters on $\mathrm{Gal}(K_\ell/K)$. One of them, which is just the restriction of χ_ℓ, is fixed under conjugation by the non-trivial element of $\mathrm{Gal}(K/\mathbf{Q})$; the other, which will be denoted by η_ℓ, is taken into η^{-1} by such conjugation. But conjugation of $\rho(\sigma)$ with a suitable element of $\rho(\mathrm{Gal}(K_\ell/\mathbf{Q}))$, to wit the matrix (9) with $z = -1$, simply changes the signs of $b(\sigma)$ and $c(\sigma)$; so it reverses the sign of the homomorphism (41) and that homomorphism therefore factors through $\log \eta$.

Explicitly we must have

$$f(\sigma) \equiv \theta \log \eta(\sigma) \mod \ell^2 \tag{43}$$

for some θ in $\mathbf{Q}_\ell^*$. But since the two summands in (40) are congruent mod ℓ, their product is congruent to $(\frac{1}{2}f(\sigma))^2 \mod \ell^2$; in other words

$$\ell^{-N}b(\sigma)c(\sigma) \equiv \tfrac{1}{4}\lambda^{-1}\chi^{k-1}(\sigma)f^2(\sigma) \mod \ell^2.$$

Moreover

$$\chi^{-\mu}(a-\chi^\mu)(d-\chi^\mu) + \chi^{\mu+1-k}(a-\chi^{k-1-\mu})(d-\chi^{k-1-\mu}) \equiv 0 \mod \ell^{N+2}$$

by (29) again, with $N > 1$. This implies

$$\begin{aligned} 2(a+d) &\equiv (\chi^{-\mu}+\chi^{\mu+1-k})(ad+\chi^{k-1}) \\ &\equiv (\chi^{k-1-\mu}+\chi^\mu)(2+\chi^{1-k}bc) \mod \ell^{N+2}; \end{aligned}$$

taking σ = Frob(p) with $(\frac{p}{\ell}) = +1$ we finally obtain

$$\tau(p) = a + d \equiv (p^{k-1-\mu}+p^\mu)(1+\tfrac{1}{8}\ell^N\lambda^{-1}f^2(\sigma)) \mod \ell^{N+2}.$$

Suppose in particular that ℓ = 3 and that u,v are given by (35); then after replacing η by a certain power of itself we can assume

$$\eta(\text{Frob } \pi) = (u+3v\sqrt{-3})/(u-3v\sqrt{-3}),$$

for this clearly has the right behaviour under Gal(K/**Q**). It follows from this and (43) that f(Frob π) is congruent mod 3^2 to a constant multiple of v/u. We can sum up what we have proved as follows :

THEOREM 6. *Suppose that* ℓ = 3 *and* k *is such that* $N > 1$; *then there exists* κ *in* $\mathbf{Z}_3^*$, *depending only on* k, *such that* $4p = u^2 + 27v^2$ *implies*

$$\tau(p) \equiv (p^{k-1-\mu}+p^\mu)(1+3^N\kappa v^2/u^2) \mod 3^{N+2}.$$

In this congruence μ *can be replaced by any integer* $m \equiv \mu \mod 3^N$.

For the six known cases of this Theorem we have the following table, which repeats some of the information in the table which follows (36).

k	12	16	18	20	22	26
N	6	5	5	5	6	5
$\mu \bmod 3^N$	119	174	143	55	18	97
κ	7	1	8	8	5	8

It is easily checked that the result for $k = 12$ is equivalent to (37); and there are analogous simplifications of the other five results.

To prove Atkin's conjectured congruence (38) we need a substitute for Lemma 4 valid when $N = 1$; but for the applications we can assume $\ell > 3$, which is a substantial simplification.

LEMMA 5. *Suppose that we are in case (c), and that $\ell > 3$ and $N = 1$; and denote by λ any pull-back to $\mathbf{Z}$ of the λ in (29). Then the map*

$$\sigma \mapsto (b\chi^{-\mu}+\lambda\ell^{-1}c\chi^{\mu+1-k})(1-\tfrac{1}{6}bc\chi^{1-k}) \tag{44}$$

induces a homomorphism

$$\mathrm{Gal}(K_\ell/K) \to \mathbf{Z}/(\ell^2). \tag{45}$$

PROOF. Denote the map (44) by g and the map (40) as before by f; and adopt the notation of the proof of Lemma 4 with the additional convention that $n_i = n(\sigma_i)$, where n is as in (29). The first displayed formula in the proof of Lemma 4 is still valid, but we now have to use the congruences

$$a\chi^{-\mu} \equiv d\chi^{\mu+1-k} \equiv 1 + \tfrac{1}{2}\lambda n^2\ell \mod \ell^2$$

instead of (42). This gives, mod ℓ^2,

$$f_1(\sigma_1\sigma_2) - f(\sigma_1) - f(\sigma_2) \equiv \tfrac{1}{2}\lambda n_1^2(\ell b_2\chi_2^{-\mu}+\lambda c_2\chi_2^{\mu+1-k})$$

$$+ \tfrac{1}{2}\lambda n_2^2(\ell b_1\chi_1^{-\mu}\chi_2^{k-1-2\mu}+\lambda c_1\chi_1^{\mu+1-k}\chi_2^{2\mu+1-k})$$

$$\equiv \lambda^2\ell n_1^2 n_2 + \lambda^2\ell n_2^2 n_1 \tag{46}$$

by a further use of (29) for b and c. But (29) also gives

$$g(\sigma) - f(\sigma) \quad -\tfrac{1}{3}\lambda^2\ell n^3 \quad \bmod \ell^2,$$

and the map $\sigma \mapsto n \bmod \ell$ is a homomorphism; thus if we write $g - f = h$ we have

$$h(\sigma_1\sigma_2) - h(\sigma_1) - h(\sigma_2) \equiv -\lambda^2\ell n_1 n_2(n_1+n_2) \quad \bmod \ell^2.$$

The combination of this and (46) proves the Lemma.

The homomorphism (45), like (41), has its sign reversed by conjugation with the non-trivial element of $\mathrm{Gal}(K/\mathbf{Q})$; so an argument exactly like the proof of (43) now gives

$$g(\sigma) \equiv \theta \log \eta(\sigma) \quad \bmod \ell^2 \tag{47}$$

for some θ in $\mathbf{Q}_\ell^*$. Since the two summands in the first factor on the right in (44) are congruent mod ℓ, we can again write down their product mod ℓ^2 and thereby obtain

$$\lambda\ell^{-1}\chi^{1-k}bc \equiv \tfrac{1}{4}g^2(1-\tfrac{1}{6}bc\chi^{1-k})^{-2} \quad \bmod \ell^2$$

from which it follows that

$$\ell^{-1}bc \equiv \tfrac{1}{4}\lambda^{-1}\chi^{k-1}g^2(1+\ell\lambda^{-1}g^2/12) \quad \bmod \ell^2.$$

Moreover (29) in this case gives, mod ℓ^3,

$$\chi^{-\mu}(a-\chi^{\mu})(d-\chi^{\mu}) + \chi^{\mu+1-k}(a-\chi^{k-1-\mu})(d-\chi^{k-1-\mu})$$

$$\equiv \tfrac{1}{4}\lambda^2 n^4\ell^2(\chi^{k-1-\mu}+\chi^{\mu}) \equiv 2^{-6}\lambda^{-2}\ell^2(a+d)g^4$$

which implies

$$(2+2^{-6}\lambda^{-2}\ell^2 g^4)(a+d) \equiv (\chi^{-\mu}+\chi^{\mu+1-k})(ad+\chi^{k-1})$$

$$\equiv (\chi^{k-1-\mu}+\chi^{\mu})(2+\chi^{1-k}bc) \quad \mod \ell^3.$$

Taking σ = Frob(p) with $(\frac{p}{\ell}) = +1$ we finally obtain

$$\tau(p) = a + d \equiv (p^{k-1-\mu}+p^{\mu})(1+\tfrac{1}{8}\lambda^{-1}\ell g^2+\lambda^{-2}\ell^2 g^4/384) \quad \mod \ell^3,$$

where g is given by (47).

The five cases we are concerned with all have $\ell = 7$ or 11; and if $(\frac{p}{\ell}) = +1$ we can write

$$4p = u^2 + \ell v^2 = 4\pi\bar{\pi} \qquad (\ell = 7 \text{ or } 11)$$

in essentially only one way, where $2\pi = u+v\sqrt{-\ell}$. We can assume

$$\eta(\text{Frob } \pi) = (u+v\sqrt{-\ell})/(u-v\sqrt{-\ell}),$$

since this clearly has the right behaviour under Gal(K/**Q**). It follows from this and (47) that g(Frob π) is congruent mod ℓ^2 to a constant multiple of

$$v/u - \ell v^3/3u^3.$$

Cleaning up, we obtain the following result

THEOREM 7. *Suppose that* $\ell = 7$ *or* 11, *and that* k *is such that we are in case (c) with* N = 1. *Then there exists* κ *in* $\mathbf{Z}_\ell^*$, *depending only on* k, *such that* $4p = u^2 + \ell v^2$ *implies*

$$\tau(p) \equiv (p^{k-1-\mu}+p^{\mu})(1+6\kappa\ell v^2/u^2+\kappa(6\kappa-4)\ell^2 v^4/u^4) \quad \mod \ell^3.$$

In this congruence μ *can be replaced by any integer* $m \equiv \mu \mod \ell(\ell-1)$.

This should of course be read with Theorem 5. For the five known cases of this Theorem we have the following table.

k	12	18	20	26	18
ℓ	7	7	7	7	11
$\mu \bmod \ell(\ell-1)$	1	1	2	2	1
κ	13	11	40	1	35

The values in the first column imply (38) and indeed give a stronger result. The fact that in each column of this table we can choose m to be so small is almost certainly significant - particularly since the machinery of [3] virtually forces the values of m in that paper to be very small; but I can see no explanation for it.

REFERENCES

[1] N.M. KATZ, Higher congruences between modular forms, Ann. Math. (2) 101 (1975), 332-367.

[2] K.A. RIBET, On ℓ-adic representations attached to modular forms, Inventiones Math. 28 (1975), 245-275.

[3] H.P.F. SWINNERTON-DYER, On ℓ-adic representations and congruences for coefficients of modular forms, in Modular Functions of One Variable III (Springer Lecture Notes, vol. 350), 1-55.

H.P.F. Swinnerton-Dyer
Department of Mathematics
Cambridge University
16 Mill Lane
Cambridge CB2 1SB, England

ON SOME CONGRUENCES BETWEEN CUSP FORMS ON $\Gamma_0(N)$.*)

By Koji Doi and Masami Ohta

---

*) Some of the examples in this paper were discussed by the first-named author at the Bonn Conference under the title "Some problems on Fourier coefficients of modular forms of one variable". He was supported by the Sonderforschungsbereich "Theoretische Mathematik", University of Bonn, during the year 1976/77.

INTRODUCTION

Let N be a prime number, and $\Gamma_o(N)$ be the group:

$$\Gamma_o(N) = \left\{ \begin{pmatrix} a & b \\ c & d \end{pmatrix} \in SL_2(\mathbb{Z}) \;\middle|\; c \equiv 0 \bmod N \right\}.$$

This group acts on the complex upper half plane $H = \{ z \in \mathbb{C} \mid \operatorname{Im}(z) > 0 \}$. We denote by $X_o(N)$ the "canonical model" of the natural compactification of $H/\Gamma_o(N)$, which is defined over $\mathbb{Q}$, and by $J_o(N)$ the jacobian variety of $X_o(N)$ defined over $\mathbb{Q}$. Let $g_o(N)$ be the genus of $X_o(N)$. We denote by $S_2(\Gamma_o(N))$ the space of all cusp forms of weight 2 with respect to $\Gamma_o(N)$, and write

$$S_2^+(\Gamma_o(N)) = \left\{ f \in S_2(\Gamma_o(N)) \;\middle|\; f\left|\begin{pmatrix} 0 & -1 \\ N & 0 \end{pmatrix}\right. = f \right\},$$

$$S_2^-(\Gamma_o(N)) = \left\{ f \in S_2(\Gamma_o(N)) \;\middle|\; f\left|\begin{pmatrix} 0 & -1 \\ N & 0 \end{pmatrix}\right. = -f \right\}.$$

Let $f(z)$ be an element of $S_2(\Gamma_o(N))$, and let $f(z) = \sum_{n=1}^{\infty} a_n q^n$ be its Fourier expansion ($q = e^{2\pi i z}$). We assume that $f(z)$ is a common eigenfunction of all the Hecke operators, and that $a_1 = 1$. We denote by K_f the field generated by all a_n over $\mathbb{Q}$, and by Σ_f the set of all the distinct embeddings of K_f into $\mathbb{C}$. For a prime number p, we denote by $H_{f,p}$ the "p-th Hecke polynomial" of f:

$$H_{f,p}(T) = \prod_{\sigma \in \Sigma_f} (T - a_p^{\sigma}),$$

where T is an indeterminate. This is a polynomial of degree $[K_f : \mathbb{Q}]$ with rational integral coefficients. We will also consider a second eigenfunction $g(z) = \sum_{n=1}^{\infty} b_n q^n$ $(b_1 = 1)$ with the obvious definitions of K_g, Σ_g, and $H_{g,p}(T)$; it will be always assumed that $g(z)$ is not a "companion" of $f(z)$, i.e., there is no element $\sigma \in \Sigma_f$ which satisfies $a_n^{\sigma} = b_n$ for all n.

The purpose of this paper is firstly to report that $H_{f,p}(T) \bmod \ell$ and $H_{g,p}(T) \bmod \ell$, which are considered as polynomials with coefficients in $\mathbb{Z}/\ell\mathbb{Z}$, have non-trivial common factors for some prime ℓ and for small p. The numerical table is given in §1. We shall then study the meaning of the existence of such a congruence. Let R_1 (resp. R_2) be the subring of K_f (resp. K_g) which is generated by all a_n (resp. b_n) over $\mathbb{Z}$. Then in some cases, we can prove a stronger congruence:

(*) There is a maximal ideal $\mathfrak{l}_i$ of R_i $(i = 1, 2)$ whose residue characteristic is ℓ such that

(1) $R_1/\mathfrak{l}_1$ and $R_2/\mathfrak{l}_2$ are isomorphic,

(2) By identifying $R_1/\mathfrak{l}_1$ with $R_2/\mathfrak{l}_2$ under the isomorphism in (1), we have $a_n \bmod \mathfrak{l}_1 = b_n \bmod \mathfrak{l}_2$ for all n.

This topic is studied in §2.

In §3, we shall study the relation between the congruence (*) and the structure of the Galois modules of $\mathfrak{l}_i$-section points of the abelian varieties corresponding to f and g. In §4, we shall study the structure of the ring of Hecke operators associated to the pair (f, g) in connection with (*).

§1. Congruences of Hecke polynomials; numerical table.

Let the notation be as in the introduction. For f(z) and g(z) as above, it is known that a_N (resp. b_N) is equal to either -1 or +1 according as f(z) (resp. g(z)) belongs to $S_2^+(\Gamma_0(N))$ or $S_2^-(\Gamma_0(N))$ (cf. Atkin, Lehner [1] Th.3). Therefore if $a_N \neq b_N$, $H_{f,N}(T) \bmod \ell$ and $H_{g,N}(T) \bmod \ell$ have no common factor unless $\ell = 2$. For example, for N = 37, 43, 53, 61, we can in fact prove the existence of the congruence (*) with ℓ_i dividing 2, but for N = 67, there is no congruence between $f \in S_2^+(\Gamma_0(67))$ and $g \in S_2^-(\Gamma_0(67))$ with $b_n \in \mathbb{Z}$. In the following discussions, we keep our attention only for the case $a_N = b_N$, i.e., we assume that f and g both belong to $S_2^+(\Gamma_0(N))$ or both to $S_2^-(\Gamma_0(N))$. We shall give all the possible values of ℓ such that $H_{f,p}(T) \bmod \ell$ and $H_{g,p}(T) \bmod \ell$ have non-trivial common factors for small p (at least for $p \leq 2g_0(N) - 1$) for $N \leq 223$. For the computation, we used the table of Wada [17] and the table in [18] (Table 5).

In the following table, we describe the cases where the spaces $S_2^{\pm}(\Gamma_0(N))$ have non-trivial splittings and list (in the last column) the values of ℓ for which congruences have been found. We underline the ℓ for which we could prove the stronger congruence (*) (by the method described in §2).

Table (1.1)

N	sign ε	splitting of $S_2^{\varepsilon}(\Gamma_0(N))$	possible ℓ
67	-	1 + 2	<u>5</u>
71	-	3 + 3	<u>3</u>
73	-	1 + 2	<u>3</u>

89	-	1 + 5	<u>5</u>
109	-	1 + 4	<u>2</u>
113	-	1 + 2 + 3	<u>2</u> for "1 + 2" <u>3</u> for "1 + 3" 11 for "2 + 3"
139	-	1 + 7	<u>2</u>, <u>3</u>
151	-	3 + 6	2, 67
163	+	1 + 5	<u>3</u>
179	-	1 + 11	<u>3</u>
193	+	2 + 5	11
197	+	1 + 5	<u>5</u>
199	-	2 + 10	71
211	+	3 + 3	<u>7</u>
	-	2 + 9	41
223	+	2 + 4	7

<u>Remark</u> (1.2) As we have remarked above, except for the underlined ℓ, it is not known to us whether $H_{f,p}(T) \bmod \ell$ and $H_{g,p}(T) \bmod \ell$ have common factors for all p. But it should be noted that the set of "possible ℓ" is non-empty whenever the spaces $S_2^{\varepsilon}(\Gamma_0(N))$ decompose, within the limit of the table.

§2. <u>Congruences of cusp forms</u>.

Let f(z) and g(z) be as in the introduction. We use the following lemma to deduce the congruence (*) from the congruence for a finite number of Hecke polynomials.

Lemma (2.1). Let K be an algebraic number field of finite degree which contains K_f and K_g. Let $\mathfrak{l}$ be a prime ideal of K which does not divide the level N. If $a_p \bmod \mathfrak{l} = b_p \bmod \mathfrak{l}$ for all prime numbers p such that $p \leq 2g_o(N) - 1$, then we have $a_n \bmod \mathfrak{l} = b_n \bmod \mathfrak{l}$ for all n.

Proof. First note that $X_o(N) \otimes K$ has good reduction mod $\mathfrak{l}$ by Igusa [6]. We also remark the following points:

(1) By the natural isomorphism: $S_2(\Gamma_o(N)) \xrightarrow{\sim} H^0(X_o(N) \otimes C, \Omega^1)$ which sends h to $h \frac{dq}{q}$, $f \frac{dq}{q}$ and $g \frac{dq}{q}$ are differential forms of the first kind rational over K.

(2) The cusp at infinity determines a $\mathbb{Q}$-rational point of $X_o(N)$, and q is a local parameter at this point. q has the same property in characteristic ℓ (ℓ being the residue characteristic of $\mathfrak{l}$).

(3) The expansion of $(f \frac{dq}{q}) \bmod \mathfrak{l}$ at infinity is given by $\sum_{n=1}^{\infty} (a_n \bmod \mathfrak{l}) q^{n-1} dq$, and the same holds for g.

Then by the Riemann-Roch theorem, if $a_n \bmod \mathfrak{l} = b_n \bmod \mathfrak{l}$ for all $n \leq 2g_o(N) - 1$, we conclude that $(f \frac{dq}{q}) \bmod \mathfrak{l} = (g \frac{dq}{q}) \bmod \mathfrak{l}$. But since f and g are common eigenfunctions of all the Hecke operators, this condition is satisfied under our assumption. q.e.d.

We give here two typical examples. The other cases, which are underlined in the table (1.1), can be verified by a similar method.

Example 1. N = 73. We consider the space $S_2^-(\Gamma_o(73))$. In this case, we can take f(z) (resp. g(z)) so that $a_n \in \mathbb{Z}$ (resp. $b_n \in \mathbb{Z}[\frac{1+\sqrt{13}}{2}]$) for all n (and in fact $R_2 \cong \mathbb{Z}[\frac{1+\sqrt{13}}{2}]$). By a result of Koike [7], we can determine the values of b_n.

p	2	3	5	7
a_p	1	0	2	2
b_p	$\frac{1-\sqrt{13}}{2}$	$\frac{1+\sqrt{13}}{2}$	$\frac{-1+\sqrt{13}}{2}$	-1

($1 - b_p + p \equiv 0 \bmod (4 - \sqrt{13})$ for all primes $p \neq 73$.)

Thus we see that $a_p \bmod 3 = b_p \bmod (4 + \sqrt{13})$ for $p \leq 7$. Hence we have $a_n \bmod 3 = b_n \bmod (4 + \sqrt{13})$ for all n, by the Lemma (2.1).

Example 2. $N = 89$. We consider the space $S_2^-(\Gamma_0(89))$ which splits as "1 + 5". We can take f(z) (resp. g(z)) so that $a_n \in \mathbb{Z}$ (resp. $b_n \in K_g$ with $[K_g : \mathbb{Q}] = 5$) for all n. By the table of Wada [17], we have:

$$H_{g,3}(T) = T^5 + 3T^4 - 4T^3 - 16T^2 - 9T - 1,$$

which is irreducible over $\mathbb{Q}$. Therefore b_3 generates K_g over $\mathbb{Q}$. By an easy computation, we have:

(1) The discriminant of the order $\mathbb{Z}[b_3] \subset K_g$ is equal to $2^4 \cdot 5 \cdot 6689$.

(2) $H_{g,3}(T) \equiv (T - 2)^2 (T^3 + 2T^2 + 1) \bmod 5$, where the last factor of the right hand side is irreducible over $\mathbb{Z}/5\mathbb{Z}$.

Then we see that 5 is prime to the conductor of R_2, and 5 decomposes in K_g as $\mathfrak{l}_2^2 \mathfrak{l}_2'$ where $\mathfrak{l}_2$ and $\mathfrak{l}_2'$ are primes of K_g, and their absolute norms are equal to 5 and 5^3, respectively. On the other hand, one can check that $H_{g,p}(T) \bmod 5 = ((T - a_p) \bmod 5)^2 I_p(T)$ with a $\mathbb{Z}/5\mathbb{Z}$-irreducible $I_p(T)$ for $p \leq 13$. We then conclude that $a_n \bmod 5 = b_n \bmod \mathfrak{l}_2$ for all n, by the Lemma (2.1).

§3. The Galois modules of ℓ_i-section points of abelian varieties.

The discussions of this and the next section are based on the existence of the congruence (*). As above, we start with f and g, and we assume that f, g, and ℓ_i (i = 1, 2) satisfy (*). In this section, we also assume the following

(3.1). ℓ_i is prime to the conductor of R_i (i = 1, 2).

By Shimura [15], we obtain an abelian variety A_f (resp. A_g) associated to f (resp. g) as a factor of $J_o(N)$. A_f (resp. A_g) is defined over $\mathbb{Q}$, and R_1 (resp. R_2) acts on this abelian variety unitarily as its $\mathbb{Q}$-endomorphisms. Let ${}_{\ell_1}A_f(\bar{\mathbb{Q}})$ (resp. ${}_{\ell_2}A_g(\bar{\mathbb{Q}})$) be the group of ℓ_1- (resp. ℓ_2-) section points of A_f (resp. A_g) in the sense of Shimura, Taniyama [16]. Under the assumption (3.1), ${}_{\ell_1}A_f(\bar{\mathbb{Q}})$ (resp. ${}_{\ell_2}A_g(\bar{\mathbb{Q}})$) is isomorphic to $(R_1/\ell_1)^2$ (resp. $(R_2/\ell_2)^2$) as R_1- (resp. R_2-) modules. Therefore we obtain natural representations of the Galois group:

$$(3.2)\begin{cases} \rho_{f,\ell_1}: \mathrm{Gal}(\bar{\mathbb{Q}}/\mathbb{Q}) \longrightarrow \mathrm{Aut}_{R_1}({}_{\ell_1}A_f(\bar{\mathbb{Q}})) \cong GL_2(R_1/\ell_1), \\ \rho_{g,\ell_2}: \mathrm{Gal}(\bar{\mathbb{Q}}/\mathbb{Q}) \longrightarrow \mathrm{Aut}_{R_2}({}_{\ell_2}A_g(\bar{\mathbb{Q}})) \cong GL_2(R_2/\ell_2). \end{cases}$$

By the Eichler-Shimura congruence relation (cf. Shimura [14] 7.4, 7.5), we have:

(3.3) Let F_p be a Frobenius element in $\mathrm{Gal}(\bar{\mathbb{Q}}/\mathbb{Q})$ with a prime $p \neq \ell$, N. Then under the assumption (3.1), the congruence (*) implies that the characteristic polynomials of $\rho_{f,\ell_1}(F_p)$ and $\rho_{g,\ell_2}(F_p)$ coincide.

This, combined with the density theorem of Čebotarev, implies that the representations ρ_{f,ℓ_1} and ρ_{g,ℓ_2} are equivalent provided that these representations are semi-simple. For this, we have the following

Proposition (3.4).*) Let the notation be as above (and N be prime as usual), and assume that the condition (3.1) is satisfied for ℓ_1. If the representation ρ_{f,ℓ_1} is reducible, then we have: $\ell_1 \ni 1 - a_p + p$ for all primes $p \neq N$.

Proof. Let $A_f/\mathbb{Z}$ be the Néron model of A_f over $\mathbb{Z}$. We denote by G the group of ℓ_1-section points of $A_f/\mathbb{Z}$ in the schemetic sense (cf. Giraud [3]), i.e. G represents the functor: $T \mapsto \mathrm{Hom}_{R_1}(R_1/\ell_1, A_f/\mathbb{Z}(T))$ for ($\mathbb{Z}$-) schemes T. Since N is prime, A_f has semi-stable reduction at N (cf. Deligne, Rapoport [2] VI 6.9), and has good reduction outside N ([6]). From this, one sees that G is a quasi-finite flat group scheme over $\mathbb{Z}$, and $G \otimes \mathbb{Z}[\frac{1}{N}]$ is finite and of rank $\#(R_1/\ell_1)^2$ over $\mathbb{Z}[\frac{1}{N}]$. Moreover, G is a scheme in R_1/ℓ_1-vector spaces (schéma en R_1/ℓ_1-vectoriels) in the sense of Raynaud [11].

Assume that there exists a $\mathrm{Gal}(\bar{\mathbb{Q}}/\mathbb{Q})$-stable R_1/ℓ_1-subspace of dimension one in $G(\bar{\mathbb{Q}})$. We then first claim that there exists a subscheme in R_1/ℓ_1-vector spaces H of G, which is finite, flat and of rank $\#(R_1/\ell_1)$ over $\mathbb{Z}$. In fact, if G itself is finite over $\mathbb{Z}$, this is obvious. If not, we have: $G \otimes \mathbb{Z}_N = X \amalg X'$, where X is finite over $\mathbb{Z}_N$, and $X' \otimes \mathbb{F}_N = \phi$ ([4] II (6.2.6)). In our case, X is flat and this is a scheme in R_1/ℓ_1-vector spaces of rank $\#(R_1/\ell_1)$ over $\mathbb{Z}_N$. By the universal property of the Néron model, $X(\bar{\mathbb{Q}}_N)$ is the unique non-trivial R_1/ℓ_1-subspace of $G(\bar{\mathbb{Q}}_N)$ which is stable under the action of the inertia group in $\mathrm{Gal}(\bar{\mathbb{Q}}_N/\mathbb{Q}_N)$ (cf. the arguments of Grothendieck [5] 2.2.5 and 5.7). Therefore in particular $G \otimes \mathbb{Q}$ has the unique subscheme in R_1/ℓ_1-vector spaces $H_{\mathbb{Q}}$ which is of rank $\#(R_1/\ell_1)$ over $\mathbb{Q}$. The schemetic closure H of $H_{\mathbb{Q}}$ in G then satisfies the

---

*) This might be known to specialists; cf. Mazur, Serre [8], and also Ribet [13].

desired property.

It is known that such an H must be isomorphic to the constant scheme or its Cartier dual ([10]§1). Therefore, by the Eichler-Shimura congruence relation, $1 - a_p + p$ annihilates $H \otimes \mathbb{F}_p$ $(p \nmid N)$. But if $1 - a_p + p$ is not contained in $\mathfrak{l}_1$, this induces an automorphism of H over $\mathbb{Z}$, which is impossible. q.e.d.

Now for the underlined ℓ in the table (1.1), and for the corresponding $\mathfrak{l}_i$ $(i = 1, 2)$ which satisfy (*), we can check (3.1). Except for the "1 + 2" part of $N = 113$, we can also check that $\mathfrak{l}_1$ (resp. $\mathfrak{l}_2$) does not contain $1 - a_p + p$ (resp. $1 - b_p + p$) for some prime $p \nmid N$. This implies that $\rho_{f,\mathfrak{l}_1}$ and $\rho_{g,\mathfrak{l}_2}$ are equivalent in such cases, by the above proposition. We give again two examples.

<u>Example 1 (bis)</u>. Let the situation be as in Example 1 in §2. Then $R_1 \cong \mathbb{Z}$ and $R_2 \cong \mathbb{Z}[\frac{1+\sqrt{13}}{2}]$. We have:

$$\begin{cases} 1 - a_2 + 2 = 2, \text{ which is prime to } \mathfrak{l}_1 = (3), \\ 1 - b_2 + 2 = \frac{5+\sqrt{13}}{2}, \text{ which is prime to } \mathfrak{l}_2 = (4+\sqrt{13}). \end{cases}$$

<u>Example 3.</u> $N = 71$. We consider the space $S_2^-(\Gamma_0(71))$ which splits as "3 + 3". In this case, we have:

$$\begin{cases} H_{f,2}(T) = T^3 - 5T + 3, \\ H_{g,2}(T) = T^3 + T^2 - 4T - 3. \end{cases}$$

The discriminant of the orders $\mathbb{Z}[a_2]$ and $\mathbb{Z}[b_2]$ are both prime to 3, and 3 decomposes in K_f (resp. K_g) as follows:

$$\begin{cases} (3) = \mathfrak{l}_1 \mathfrak{l}_1' \text{ in } K_f, \\ (3) = \mathfrak{l}_2 \mathfrak{l}_2' \text{ in } K_g, \end{cases}$$

where the absolute norm of $\mathfrak{l}_1$ (resp. $\mathfrak{l}_2$) is equal to 3, and that of $\mathfrak{l}_1'$ (resp.

$\mathfrak{l}_2'$) is equal to 9. Also we know that $a_n \bmod \mathfrak{l}_1 = b_n \bmod \mathfrak{l}_2$ for all n, and there is no other congruence. On the other hand, we have:

$$\begin{cases} H_{f,5}(T) = T^3 + 3T^2 - 2T - 7, \\ H_{g,5}(T) = T^3 - 5T^2 - 2T + 25. \end{cases}$$

The absolute norm of $1 - a_5 + 5$ (resp. $1 - b_5 + 5$) is obviously prime to 3.

§4. The Hecke ring associated to the pair (f, g).

Let f and g be as before, and assume the congruence (*) for f, g and $\mathfrak{l}_i$ ($i = 1, 2$). We denote by R the subring of $R^* = R_1 \oplus R_2$ which is generated over $\mathbb{Z}$ by all $(a_n, b_n) \in R^*$. The purpose of this section is to study the relation between (*) and the structure of R.

We begin with elementary ring theory. In general, let R_i ($i = 1, 2$) be an integral domain, and R be a subring of $R^* = R_1 \oplus R_2$ such that $[R^* : R] < \infty$. We denote by $\mathfrak{c}$ the conductor of R in R^*:

(4.1) $\quad \mathfrak{c} = R : R^* = \{ r \in R^* \mid rR^* \subseteq R \}$.

$\mathfrak{c}$ is an ideal of R^* which is contained in R. Obviously we have:

(4.2) $\quad \mathfrak{c} = \mathfrak{c}_1 \oplus \mathfrak{c}_2$ with $\mathfrak{c}_i$ an ideal of R_i ($i = 1, 2$).

Assume that the natural projection $p_i : R \longrightarrow R_i$ is surjective ($i = 1, 2$). Then, in this case, we easily see:

(4.3) The natural projection p_i induces an isomorphism:

$$R/\mathfrak{c} \longrightarrow R_i/\mathfrak{c}_i \quad (i = 1, 2).$$

Therefore for a maximal ideal $\mathfrak{l}$ of R which contains $\mathfrak{c}$, we have an isomorphism: $R/\mathfrak{l} \longrightarrow R_i/\mathfrak{l}_i$ ($i = 1, 2$) via p_i, where $\mathfrak{l}_i = p_i(\mathfrak{l})$.

Conversely, let $\mathfrak{l}_i$ be a maximal ideal of R_i ($i = 1, 2$) which satisfies

(4.4). There is an isomorphism $\varphi : R_1/\mathfrak{l}_1 \longrightarrow R_2/\mathfrak{l}_2$ such that $\varphi . q_1 = q_2$, where q_i is the natural homomorphism: $R \longrightarrow R_i/\mathfrak{l}_i$.

Then it is also easy to see that $\mathfrak{l}$ = Ker q_1 = Ker q_2 is a maximal ideal of R which contains $\mathfrak{c}$.

Now this situation can be applied to our previous R_i, R^* and R. Summing up what we have said, we obtain the follwing

Proposition (4.5). The congruence (*) holds for $\mathfrak{l}_i$ (i = 1, 2) if and only if $(\mathfrak{l}_1 \oplus \mathfrak{l}_2) \cap R$ is a maximal ideal of R which contains $\mathfrak{c}$.

Let us consider the underlined cases in the table (1.1).

Example 4. N = 139. We consider the space $S_2^-(\Gamma_0(139))$ which splits as "1 + 7". Here we have the congruence (*) with $\mathfrak{l}_i$ (i = 1, 2) dividing 2, and with $\mathfrak{l}_i'$ (i = 1, 2) dividing 3. On the other hand, by [17] , we have:

$$\begin{cases} H_{f,2}(T) = T - 1, \\ H_{g,2}(T) = T^7 - T^6 - 11T^5 + 8T^4 + 35T^3 - 10T^2 - 32T - 8, \end{cases}$$

$$\begin{cases} H_{f,5}(T) = T + 1, \\ H_{g,5}(T) = T^7 - 11T^6 + 36T^5 + 2T^4 - 211T^3 + 319T^2 - 55T - 83. \end{cases}$$

From this, we see that $(0, 1 - b_2)$, $(0, -1 - b_5) \in R$, and the absolute norm of $1 - b_2$ (resp. $-1 - b_5$) is equal to -18 (resp. 456). Therefore we have:

$$\begin{cases} R = \{(a, b) \in R^* \mid a \bmod \mathfrak{l}_1 = b \bmod \mathfrak{l}_2,\ a \bmod \mathfrak{l}_1' = b \bmod \mathfrak{l}_2'\}, \\ \mathfrak{c} = (\mathfrak{l}_1 \oplus \mathfrak{l}_2) \cap (\mathfrak{l}_1' \oplus \mathfrak{l}_2'). \end{cases}$$

Except for this, and the case N = 71 and N = 109, we can check that there is an element $(0, b) \in R$ such that the absolute norm of b is not divisible by

ℓ^2. Therefore for these ten cases, we have:

$$\begin{cases} R = \{(a,\ b) \in R^* \mid a \bmod \mathfrak{l}_1 = b \bmod \mathfrak{l}_2\}, \\ \mathfrak{L} = \mathfrak{l}_1 \oplus \mathfrak{l}_2. \end{cases}$$

Concluding remark. Let K_f and K_g be as before. Since f and g are not companions to each other, $K_f \oplus K_g$ can be considered as a direct factor of $\mathrm{End}_{\mathbb{Q}}(J_o(N)) \otimes \mathbb{Q}$, which is isomorphic to a sum of totally real fields. Then by the same method as in Shimura [15], we obtain an abelian variety $A_{(f,g)}$ which is the factor (as a quotient) of $J_o(N)$ corresponding to $K_f \oplus K_g$. $A_{(f,g)}$ is a quotient of $J_o(N)$ by an abelian subvariety rational over $\mathbb{Q}$. By the construction (loc. cit.), the ring R acts on $A_{(f,g)}$ as $\mathbb{Q}$-endomorphisms, i.e.,

(4.7) There is a natural injective homomorphism

$$\theta : R \longrightarrow \mathrm{End}_{\mathbb{Q}}(A_{(f,g)}).$$

By a theorem of Ribet [12], $\theta(R)$ is of finite index in $\mathrm{End}_{\mathbb{Q}}(A_{(f,g)}) = \mathrm{End}_{\overline{\mathbb{Q}}}(A_{(f,g)})$. In [9] (§5, Prop.4), Mazur and Swinnerton-Dyer have proved that θ is surjective when $N = 37$. (In this case, $f \in S_2^+(\Gamma_o(37))$, $g \in S_2^-(\Gamma_o(37))$, $A_{(f,g)} = J_o(37)$, and $R \cong \{(a,\ b) \in \mathbb{Z} \oplus \mathbb{Z} \mid a \equiv b \bmod 2\}$.) Although we know no other example, it seems interesting to study whether θ is surjective in general or not.

References

[1] A.O.L. Atkin, J. Lehner, Hecke operators on $\Gamma_o(m)$, Math. Ann. 185, 134-160 (1970).

[2] P. Deligne, M. Rapoport, Les schémas de modules de courbes elliptipues.

in Modular functions of one variable II, Lecture Notes in Math. 349, Springer.

[3] J. Giraud, Remarque sur une formule de Shimura-Taniyama, Inv. math., 5, 231-236 (1968).

[4] A. Grothendieck (with J. Dieudonné), Eléments de géométrie algébrique , Publ. Math. I.H.E.S. 4, 8, 11, etc.

[5] A. Grothendieck, Modèles de Néron et monodromie (exposé IX de SGA 7), in Lecture Notes in Math. 288, Springer.

[6] J. Igusa, Kroneckerian model of fields of elliptic modular functions, Amer. J. Math., 81, 561-577 (1959).

[7] M. Koike, Congruences between Eisenstein series and cusp forms, Japan-U.S. Seminar on number theory, Ann Arbor, Michigan, 1975.

[8] B. Mazur, J.-P. Serre, Points rationnels des courbes modulaires $X_0(N)$, Sém. Bourbaki, n° 469, 1974/1975.

[9] B. Mazur, P. Swinnerton-Dyer, Arithmetic of Weil curves, Inv. math. , 25, 1-61 (1974).

[10] M. Ohta, The representation of Galois group attached to certain finite group schemes, and its application to Shimura's theory, to appear.

[11] M. Raynaud, Schémas en groupes de type (p,..,p), Bull. Soc. math. France, 102, 241-280 (1974).

[12] K. Ribet, Endomorphisms of semi-stable abelian varieties over number fields, Ann. of Math., 101, 555-562 (1975).

[13] K. Ribet, Division points of $J^*(N)$, Japan-U.S. Seminar on number theory, Ann Arbor, Michigan, 1975.

[14] G. Shimura, Introduction to the arithmetic theory of automorphic functions, Publ. Math. Soc. Japan, No. 11, Iwanami Shoten and Princeton University Press, 1971.

[15] G. Shimura, On the factors of the jacobian variety of a modular function field, J. of Math. Soc. Japan, 25, 523-544 (1973).

[16] G. Shimura, Y. Taniyama, Complex multiplication of abelian varieties and its application to number theory, Publ. Math. Soc. Japan, No. 6, 1961.

[17] H. Wada, A table of Hecke operators (cf. Proc. Japan Acad., 49, 380-384 (1973)).

[18] Modular functions of one variable IV, Lecture Notes in Math. 476, Springer.

Koji Doi
Department of Mathematics
Hokkaido University
Sapporo/Japan

M. Ohta
Department of Mathematics
Kyoto University
Kyoto/Japan

RATIONAL POINTS ON MODULAR CURVES

by B. Mazur

RATIONAL POINTS ON MODULAR CURVES

§1. Introduction

In the course of preparing my lectures for this conference, I found a proof of the following theorem, conjectured by Ogg (conjecture 1 [17b]):

THEOREM 1. *Let Φ be the torsion subgroup of the Mordell-Weil group of an elliptic curve* E , *over* $\mathbb{Q}$. *Then* Φ *is isomorphic to one of the following* 15 *groups:*

$$\mathbb{Z}/m\cdot\mathbb{Z} \qquad \text{for} \quad m \leq 10 \quad \text{or} \quad m = 12$$

$$\mathbb{Z}/2\cdot\mathbb{Z} \times \mathbb{Z}/2\nu\cdot\mathbb{Z} \qquad \text{for} \quad \nu \leq 4 \quad .$$

This proof will be presented here (see also [14a]).

The above 15 groups do indeed occur, for their "associated" moduli problems are of genus 0 with known rational parametrizations. [1)]

Theorem 1 fits into a broader conjecture, attributed by Cassels ([3] p. 264 ; cf. also bibliography) to the "folklore":

---

1) The equations are collected in ([10], Table 3, page 217), and were known for the most part to Fricke. It is amusing to consider, however, that, in disguised form, some of these parametrizations may have been known far earlier than that. Griffiths pointed out to me that the data of the classical Poncelet theorem (an n-gon inscribed in one conic and circumscribed about another) provides one with an elliptic curve and a point of order n on that elliptic curve. (As was known, in effect, to Jacobi. See [7] §1 d.)

But judging from the hints given in [6], the mathematician Nicolaus Fuss (1755-1826 ; a friend and student of Euler) may have found rational parametrizations of Poncelet quadrilaterals, pentagons, hexagons, heptagons and octagons (Nova Acta Petropol. XIII 1798, which I have not been able to track down).

Conjecture A

If K is a number field, there is a positive integer $B(K)$ such that for elliptic curve E over K, the torsion subgroup of $E(K)$ (the Mordell-Weil group of E over K) is of order $\leq B(K)$.

Theorem 1 also fits into a general program:

B. Given a number field K and a subgroup H of $GL_2\hat{\mathbb{Z}} = \prod_p GL_2 \mathbb{Z}_p$ classify all elliptic curves $E_{/K}$ whose associated Galois representation on torsion points maps $Gal(\overline{K}/K)$ into $H \subset GL_2 \hat{\mathbb{Z}}$.

Concerning conjecture A one has very little general information except for the case $K = \mathbb{Q}$. For no $K \neq \mathbb{Q}$ is the conjecture proved, nor does one even possess any serious lower bounds for $B(K)$.

One has, however, partial information of two sorts. Firstly, for a given number field K, and prime number p, there is a finite power of $p, q = p^{e(p,K)}$ such that no elliptic curve defined over K possesses a K - rational point of order (divisible by) q. This follows from a more general theorem of Manin (using the theory of heights and methods of Demjanenko [13] [22b]). The exponents $e(p, K)$ have recently been made effective by Berkovic [1] (using the descent techniques of [14a]).

Secondly, there is an extremely intriguing technique of associating to a K - rational torsion point on any elliptic curve over K (which is required to be 'rigidified' over K by extra data) K - rational points of some specific algebraic curve V of genus > 1 over K. In this way one obtains uniform bounds for the order of the torsion parts of the Mordell-Weil groups of those elliptic curves over (certain fields) K which possess the requisite rigidification over K, provided $V(K)$ is finite (more precisely: the bound is in terms of the cardinality of $V(K)$).[1)]

These techniques occur in the work of Demjanenko [5] in which further claims are made which are, it seems, unjustified. See [10] for a rigorous development and broadening of these methods. For a relationship between the problem of existence of rational $2N$ - torsion in elliptic curves over K and K - rational points on the Fermat curve $X^N + Y^N = 1$ see [11]. The paper of Kubert [10] should be consulted for its close study and ingenious use of these (and other) methods to obtain a number of specific applications.

Concerning the general program (B), a theorem of Serre [22a] assures us that, if we ignore elliptic curves of complex multiplication, we may take H to be a subgroup of finite index in $GL_2 \hat{\mathbb{Z}}$. As we shall see below, a diverse range of diophantine questions are embraced by program (B) (See [14a]). Included, in particular, is the problem of classifying elliptic curves over K possessing a K - rational N-isogeny (N a given integer > 1), or equivalently, a K - rational cyclic subgroup of order N. This problem, moreover, is also

---

1) This is reminiscent of a method introduced by Hellegouarch [8] where he related the existence of a $\mathbb{Q}$ - rational point of order p^h (p a prime number > 13, $h > 1$) to the existence of systems of $(p^h - 1)/2$ rational points of an appropriate (generalized) Fermat variety: $\sum_{j=1}^{N} X_j^{p^{h-1}} = 0$ where N is an integer independent of p and h.

equivalent to the problem of determining the K - rational points of the modular curve $X_0(N)$. Although our knowledge of isogenies is not as sharp as that of rational torsion, the theory of the Eisenstein ideal provides much information when $K = \mathbb{Q}$.[1)] [Ogg and I expect to find no $\mathbb{Q}$ - rational N - isogenies when $N > 163$.] For $K \neq \mathbb{Q}$, again, very little is known. To be sure, elliptic curves possessing complex multiplication must be treated specially when studying isogenies: if $E_{/K}$ is such an elliptic curve, for any rational prime N which splits in $R = \mathrm{End}_K(E)$ $(N = \pi \cdot \pi'$ with neither π nor π' units in $R)$ multiplication by π in E provides us with a K - rational N - isogeny.[2)]

Let us say, provisionally, that an isogeny is <u>large</u> if it is an N - isogeny for an integer N such that genus $X_0(N) \geq 2$ (equivalently: $N > 21$ and $N \neq 24, 25, 27, 32, 36, 49$). It is tempting to ask

<u>Question C:</u> <u>Is it true that for a given number field K, there are only a finite number of values $j_1, j_2, \cdots, j_{C(K)}$ such that if $E_{/K}$ is an elliptic curve possessing a large K - rational isogeny, then the elliptic modular invariant $j(E) = j_m$ for some $m \leq C(K)$?</u>

It would be interesting to make empirical investigations in this area. At the moment, one lacks sufficient experience to make any conjectures for $K \neq \mathbb{Q}$.

---

1) Surveys of some of the results of this theory occur in [14b], [16], [17b] and the complete details will appear in [14a]. See also §4 below.

2) If N does not split in the complex quadratic field $K = \mathbb{Q}(\sqrt{-d})$, see §4 cor. 2 below.

Notational conventions: If X is a scheme over a base S and $T \to S$ a morphism of schemes, we shall indicate the pullback of X to T $(X \times_S T)$ by $X_{/T}$. If $T = \operatorname{Spec} A$, we may also write $X_{/A}$. The T-valued points of X we denote $X(T)$, and again if $X = \operatorname{Spec} A$ we may also denote it $X(A)$. If X is a scheme over the field of complex numbers $\mathbb{C}$, then $X_{\mathbb{C}}$ denotes the underlying analytic space.

§2. Modular curves

Let $\mathbb{H}$ be the upper half-plane regarded as homogeneous space under $PSL_2\,\mathbb{R}$ by the usual action $\begin{pmatrix} a & b \\ c & d \end{pmatrix} : z \mapsto \frac{az+b}{cz+d}$. To a point $z \in \mathbb{H}$ we may associate a lattice $\Lambda_z = \mathbb{Z} + z \cdot \mathbb{Z} \subset \mathbb{C}$ and an elliptic curve $E_z = \mathbb{C}/\Lambda_z$. The lattice does not change under modification of z by any element of $\Gamma(1) = PSL_2\,\mathbb{Z}$ and one has the one-to-one correspondences

$$\left\{\begin{matrix}\text{elliptic curves}\\ \text{over } \mathbb{C}\text{, up to}\\ \text{isomorphism}\end{matrix}\right\} \longleftrightarrow \left\{\begin{matrix}\text{lattices in } \mathbb{C}\\ \text{up to}\\ \text{homothety}\end{matrix}\right\} \longleftrightarrow \Gamma(1)\backslash\mathbb{H} \xrightarrow[\approx]{j} \mathbb{C}$$

where the analytic isomorphism j is the elliptic modular invariant.

Let E be an elliptic curve defined over a field K and N an integer ≥ 1. Let $E[N]$ denote the kernel in E of multiplication by N. If N is prime to the characteristic of K, we view $E[N]$ as $Gal(\overline{K}/K)$ - module. It is a free $\mathbb{Z}/N$ module of rank 2 and the classical e_N - pairing provides us with a canonical isomorphism

$$\wedge^2 E[N] \xrightarrow[\approx]{} \mu_N \quad (= Gal(\overline{K}/K) \text{ - module of } N\text{-th roots of } 1)\,.$$ [1)]

Thus the determinant of the representation of $Gal(\overline{K}/K)$ on $E[N]$ (viewed as a 2 - dimensional representation over $\mathbb{Z}/N$) is equal to the standard character

---

1) There are, indeed, two canonical isomorphisms, which differ by sign. A convention we make below will stipulate which of these two we are choosing but this choice is irrelevant for our considerations.

$\chi : \mathrm{Gal}(\overline{K}/K) \longrightarrow (\mathbb{Z}/N)^*$ defined by the rule $\sigma(\zeta) = \zeta^{\chi(\sigma)}$ where $\sigma \in \mathrm{Gal}(\overline{K}/K)$ and $\zeta \in \mu_N(\overline{K})$. In particular, note that the image of the determinant, $\det(\mathrm{Gal}(\overline{K}/K))$, is equal to $(\mathbb{Z}/N)^*$.

Any isomorphism $E[N] \xrightarrow{\alpha_N} \mathbb{Z}/N \times \mathbb{Z}/N$ is called a <u>level</u> <u>N - structure</u>. To a level N structure α_N we may associate an N-th root of 1, $\zeta(\alpha_N) = \alpha_N^{-1}(1,0) \wedge \alpha_N^{-1}(0,1) \in \mu_N(\overline{K})$ where the wedge denotes the e_N - pairing.

If E is defined over $\mathbb{C}$, we shall say that α_N is a <u>canonical</u> level N structure provided $\zeta(\alpha_N) = \exp(2\pi i/N)$.

If $z \in H$, the level N structure $\alpha_N : E_z[N] \xrightarrow{\approx} \mathbb{Z}/N \times \mathbb{Z}/N$ obtained by sending $a/N + z \cdot b/N$ in $E_z[N] = \frac{1}{N}\Lambda_z/\Lambda_z$ to (a,b) in $\mathbb{Z}/N \times \mathbb{Z}/N$ is seen to be canonical (which pins down our choice of sign for the e_N - pairing).

If (E, α_N) is any pair consisting of an elliptic curve $E_{/\mathbb{C}}$ and a canonical level N structure $\alpha_N : E[N] \xrightarrow{\approx} \mathbb{Z}/N \times \mathbb{Z}/N$ then (E, α_N) is isomorphic to a pair (E_z, α_N) and the set of z for which this is true forms a single orbit under

$$\Gamma(N) = \left\{ g \in \Gamma(1) \;\middle|\; g \equiv \begin{pmatrix} 1 & 0 \\ 0 & 1 \end{pmatrix} \text{ modulo } N \right\}$$

(<u>the principal congruence subgroup of level N</u>). One compactifies the open Riemann surface $_{\Gamma(N)}\backslash H$ (of iso. classes of elliptic curves over $\mathbb{C}$ with "canonical" level N structures) by adjunction of a finite set of points (cusps) which may be identified with $_{\Gamma(N)}\backslash \mathbb{P}^1(\mathbb{Q})$ to obtain the compact algebraic curve $X(N)_{/\mathbb{C}}$ (the modular curve of level N):

$$X(N)_{\mathbb{C}} = {}_{\Gamma(N)}\backslash \mathbb{H} \amalg {}_{\Gamma(N)}\backslash \mathbb{P}^1(\mathbb{Q}) .$$

See ([4] II) for an interpretation of the cusps of $X(N)$ in terms of degenerate forms of elliptic curves $(\mathbb{C}^* \times \mathbb{Z}/N)$ with canonical level N structure. The curve $X(N)$ is a Galois covering of $X(1) = \mathbb{C} \cup \infty$ (via the elliptic modular function j), with Galois group $\Gamma(1)/\Gamma(N)$ ($\cong PSL_2 \mathbb{Z}/N\mathbb{Z}$, if N is a prime). If $H \subset GL_2\, \mathbb{Z}/N\mathbb{Z}$, set $H_o = H \cap SL_2\, \mathbb{Z}/N\mathbb{Z}$, and let X_H denote the intermediate covering $_{H_o}\backslash X(N)$. A non-cuspidal point of X_H is given either by a $\mathbb{C}$ - isomorphism class of pairs consisting in an elliptic curve E and an H - orbit of level N structures on E, or equivalently, an H_o - orbit of canonical level N structures. As a curve over $\mathbb{C}$, X_H is dependent only on the subgroup H_o; however X_H admits a natural structure over the field $\mathbb{Q}(\exp(2\pi i/N))^{\det H}$, the fixed field under the action of $\det H \subset (\mathbb{Z}/N)^* = \mathrm{Gal}(\mathbb{Q}(\zeta_N)/\mathbb{Q})$, where $\det H$ denotes the image of the determinant. (See [4] IV 3.20.4)

In particular, if $\det H = (\mathbb{Z}/N)^*$, then X_H is defined over $\mathbb{Q}$.

If $H = \{\begin{pmatrix} 1 & * \\ 0 & * \end{pmatrix} \in GL_2\, \mathbb{Z}/N\}$, the standard notation for X_H is $X_1(N)$ and its non-cuspidal points correspond to elliptic curves with a chosen point of order N.

If $H = \{\begin{pmatrix} * & * \\ 0 & * \end{pmatrix} \in GL_2\, \mathbb{Z}/N\}$, the standard notation for X_H is $X_0(N)$ and its non-cuspidal points classify N - isogenies.

More systematically, if $N \geq 5$ is a prime number, any proper subgroup

$H \subset GL_2\,\mathbb{Z}/N$ is conjugate to a subgroup of one of the entries of the following table, in which $\mathcal{S}_n$ denotes the symmetric group on n letters, $\mathfrak{A}_n$ the alternating group and $\mathbb{Q}(\chi_N)$ denotes the quadratic subfield of $\mathbb{Q}(e^{2\pi i/N})$.

(N = prime ≥ 5)

H	Notation for X_N	Field of definition
(a) The Borel subgroup $\begin{pmatrix} * & * \\ 0 & * \end{pmatrix}$	$X_0(N)$	$\mathbb{Q}$
(b) The normalizer of a split Cartan $\begin{pmatrix} * & 0 \\ 0 & * \end{pmatrix} \amalg \begin{pmatrix} 0 & * \\ * & 0 \end{pmatrix}$	$X_{\text{split}}(N)$	$\mathbb{Q}$
(c) The normalizer of a nonsplit Cartan subgroup $\mathbb{F}^*_{N^2} \subset GL_2\,\mathbb{Z}/N$	$X_{\text{nonsplit}}(N)$	$\mathbb{Q}$
(d) The inverse image in $GL_2\,\mathbb{Z}/N$ of $\mathcal{S}_4 \subset PGL_2\,\mathbb{Z}/N$	$X_{\mathcal{S}_4}(N)$	$\mathbb{Q}$ if $N \equiv \pm 3 \bmod 8$ $\mathbb{Q}(\chi_N)$ if $N \not\equiv \pm 3 \bmod 8$.
(e) The inverse image in $GL_2\,\mathbb{Z}/N$ of $\mathfrak{A}_5 \subset PGL_2\,\mathbb{Z}/N$ (possible only if $N \equiv \pm 1 \bmod 5$)	$X_{\mathfrak{A}_5}(N)$	$\mathbb{Q}(\chi_N)$
(f) The inverse image in $GL_2\,\mathbb{Z}/N$ of $\mathfrak{A}_4 \subset PGL_2\,\mathbb{Z}/N$	$X_{\mathfrak{A}_4}(N)$	$\mathbb{Q}(\chi_N)$

Remarks:

1. The normalizer of Cartan subgroups.

If $N \geq 5$ is a prime number, one has these formulae for the genus of $X_{split}(N)$ and $X_{nonsplit}(N)$ respectively:

$$g_{split}(N) = \frac{11 + (N-8) \cdot N - 4\left(\frac{-3}{N}\right)}{24}$$

$$g_{nonsplit}(N) = \frac{23 + (N-10) \cdot N + 6\left(\frac{-1}{N}\right) + 4\left(\frac{-3}{N}\right)}{24} .$$

where $\left(\frac{-}{N}\right)$ denotes the Legendre symbol.

It is immediate that $X_{split}(N)$ corresponds to the problem of classifying elliptic curves endowed with an unordered pair of independent N - isogenies, and $X_{nonsplit}(N)$ corresponds to the problem of classifying an elliptic curve together with a chosen subfield of order N^2 in the endomorphism ring of $E[N]$ (equivalently: an $\mathbb{F}_{N^2}$ - vector space structure on $E[N]$ determined up to the conjugation in $\mathbb{F}_{N^2}$).

One sees easily that the curve $X_{split}(N)$ is isomorphic (over $\mathbb{Q}$) to $X_0(N^2)/w$ where w is the canonical involution (induced from $z \mapsto -1/N^2z$ in $\mathbb{H}$). In contrast, the family of curves $X_{nonsplit}(N)$ does not seem to be directly related to any of the more "familiar" modular curves.

2. N - adic points of the modular curves associated to exceptional subgroups.

Serre has proved the following local result for elliptic curves (which Ribet and I have checked remains valid for abelian varieties of arbitrary dimension):

Let K be a finite extension of $\mathbb{Q}_N$ of ramification index e. Let E be an elliptic curve over K with a semi-stable Néron model over the ring of integers $\mathcal{O}_K$. Let $r : \mathrm{Gal}(\overline{K}/K) \to \mathrm{PGL}(E[N])$ denote the projective representation associated to the action of Galois on N - division points of E. Then: if $2e < N - 1$, the image of the inertia subgroup under r contains an element of $\geq (N-1)/e$.

Using this local result, one obtains a bound c(K) such that X_H possesses no K - rational points, when H is an exceptional subgroup of $GL_2\,\mathbb{Z}/N$ and $N > c(K)$. In particular, Serre has shown that $X_{\mathfrak{S}_4}(N)$ possesses no points rational over $\mathbb{Q}_N$ if $N > 13$.

3. 'Expected' rational points on X_H.[1)]

Let N be a prime number and E be an elliptic curve with complex multiplication over $\mathbb{C}$. Let $R = \mathrm{End}_{\mathbb{C}} E$. It is well known that E[N] is a free $R/N\cdot R$ - module of rank 1.[2)] Since $R/N\cdot R$ is an $\mathbb{F}_N$ - algebra of dimension 2, there are 3 possibilities:

---

1) These are closely related to the Heegner points studied by Birch and Stephens. Compare also [14a] Ch. III §2.

2) This follows e.g. from ([12] Ch. 8, §1 Cor. 1).

(a) $R/N \cdot R$ has a nontrivial radical. In this case $R/N \cdot R = \mathbb{F}_N[\varepsilon]$, where ε is a nontrivial element in the radical, necessarily of square zero.

(b) $R/N \cdot R$ is a product of two fields (each $\approx \mathbb{F}_N$). In this case we have the equation $1 = \varepsilon_1 + \varepsilon_2$ in $R/N \cdot R$ where $\varepsilon_1, \varepsilon_2$ are a pair of orthogonal idempotents.

(c) $R/N \cdot R$ is a field.

In case (a), the kernel of ε in $E[N]$ is a subgroup of order N, and the pair $(E, \ker \varepsilon)$ determines a point $a_E \in X_0(N)$.

In case (b), the kernels of ε_1 and ε_2 in $E[N]$ provide an unordered pair of independent cyclic subgroups of order N, and thereby determines a point $a_E \in X_{split}(N)$.

In case (c) we obtain a point $a_E \in X_{nonsplit}(N)$.

Consequently, for each elliptic curve with complex multiplication E, we obtain a (noncuspidal) point $a_E = a_E(N) \in X_0(N) \amalg X_{split}(N) \amalg X_{nonsplit}(N)$ which, by the theory of complex multiplication (e.g. [12] part two Chapter 10), is defined over a subfield of index two in the ray class field of $R \otimes \mathbb{Q}$, with conductor equal to the conductor of R.

Since there are 13 complex quadratic orders R with class number 1, for each N we obtain 13 $\mathbb{Q}$-rational points $a_E(N)$ in $X_0(N) \amalg X_{split}(N) \amalg X_{nonsplit}(N)$ (the "expected" rational points).

For $R = \mathbb{Z}[\frac{1+\sqrt{-N}}{2}]$, $E = \mathbb{C}/R$, and $N = 11, 19, 43, 67, 163$, the point

$a_E(N)$ lands in $X_0(N)$. For all other cases $(N \geq 11)$ the "expected" rational points distribute themselves among X_{split} and $X_{nonsplit}$.

4. <u>The status of knowledge of $\mathbb{Q}$ - rational points of the modular curves.</u>

If N is prime, and $H \subset GL_2\ \mathbb{Z}/N$ is an entry of the above table, such that $g_H > 0$, and X_H is defined over $\mathbb{Q}$, the following facts are known:

(a) $X_H = X_0(N)$ has only a finite number of $\mathbb{Q}$ - rational points. If ν denotes the number of noncuspidal rational points on $X_0(N)$, then ν is known for all $N < 250$ except $N = 151$ and 227 and $\nu = 0$ in this range except for the following values of N:

N	11	17	19	37	43	67	(151)	163	(227)
ν	3	2	1	2	1	1	?	1	?

(besides the techniques of [14a], this tabulation makes use of calculations and results due to Atkin, Brumer and Kramer, Ogg, Parry, Tingley and Wada. See [14a]).

(b) $X_H = X_{split}(N)$ has only a finite number of $\mathbb{Q}$ - rational points if $N \neq 13$. Since $X_{split}(13)$ is of genus 3, one expects that it, too, has only a finite number of rational points. ([14a] Ch. III §6).

(c) For $X_H = X_{nonsplit}(N)$ <u>nothing</u> is known.

(d) By remark 3, $X_{\mathfrak{S}_4}(N)$ has no rational points for $N > 13$. Serre, however, has constructed a rational point on $X_{\mathfrak{S}_4}(11)$ and on $X_{\mathfrak{S}_4}(13)$ using elliptic curves with complex multiplication by $\sqrt{-3}$.

To be sure, the last two entries in the table are not defined over $\mathbb{Q}$, and therefore can have no $\mathbb{Q}$ - rational points.

§3. A proof of Theorem 1

Theorem 1 implies that if $m < \infty$ is the order of a $\mathbb{Q}$ - rational torsion point of an elliptic curve over $\mathbb{Q}$, then $m \leq 10$ or $m = 12$. This statement is equivalent to:

THEOREM 2. If the genus of $X_1(m)$ is > 0 (i.e. $m = 11$ or $m > 13$), then the only $\mathbb{Q}$ - rational points on $X_1(m)$ are the $\mathbb{Q}$ - rational cusps.

Kubert ([10] IV. 1.2) has shown that to prove Theorem 1, it suffices to prove Theorem 2 for m a prime number ≥ 23; this is what we shall do below.

The proof we shall give for Theorem 2 will in fact be valid for $m = N$, a prime number such that either $N = 11$, or $N \geq 17$ (i.e. such that the genus of $X_0(N)$ is > 0). Since it may be of interest, at least for clarity of presentation, to make essential ingredients of the proof explicit at the outset, we shall do this by axiomatizing what is needed.

We shall prove:

PROPOSITION: Let (K, N) be a pair consisting in a number field K and a prime number N, which satisfies Axioms 1, 2, and 3 below.

Then the only K - rational points of $X_1(N)$ are the K - rational cusps.[1)]

---

1) At present, I have no example of a number field K different from $\mathbb{Q}$ and a prime number N such that (K, N) satisfies these axioms. But compare the next footnote with §4 Cor. 1.

Our first axiom is simply a bound:

Axiom 1: Let $d = [K:\mathbb{Q}]$. Then $N > 1 + 3^d + 2 \cdot 3^{d/2}$.

Axiom 2: There is a nonconstant map (over $\mathbb{Q}$)

$$f : X_0(N) \to A$$

where A is an abelian variety, such that:

(i) if $\underline{0}$ and $\underline{\infty}$ are the cusps of $X_0(N)$, then $f(\underline{0}) \neq f(\underline{\infty})$.

(ii) The Mordell-Weil group $A(K)$ is finite.

Remark: If $K = \mathbb{Q}$, then Axiom 2 is satisfied when A is taken to be the Eisenstein quotient of J = the jacobian of $X_0(N)$ provided $N = 11$ or $N \geq 17$. (cf. [14a] theorem 4 of introduction and §4 below).

For our third axiom, we need a definition. Let L/K be a Galois extension such that ζ_N (a primitive N-th root of 1) lies in L and such that $V = \mathrm{Gal}(L/K(\zeta_N))$ is an abelian group killed by N.

$$V\left\{\begin{array}{l} L \\ \quad\diagdown K(\zeta_N) \\ \qquad\quad\diagdown K \end{array}\right.$$

Suppose that the natural action of $\mathrm{Gal}(K(\zeta_N)/K)$ on V (via conjugation through $\mathrm{Gal}(L/K)$) is given by multiplication by the j-th power of the standard character χ. That is

$$\tau \cdot v = \chi(\tau)^j \cdot v$$

where we recall that the standard character $\chi : \mathrm{Gal}(K(\zeta_N)/K) \to (\mathbb{Z}/N)^*$

is defined by the rule: $\zeta_N^{\chi(\tau)} = \tau \cdot \zeta_N$, for $\tau \in \mathrm{Gal}(K(\zeta_N)/K)$.

Here j may be taken to be an integer modulo $N - 1$.

If the above is the case, let us refer to L as a χ^j - <u>extension</u> of $K(\zeta_N)$.

<u>Axiom 3: There are no nontrivial everywhere unramified</u> χ^{-1} - <u>extensions of</u> $K(\zeta_N)$.

<u>Remark:</u> When $K = \mathbb{Q}$, Axiom 3 is indeed valid, and follows from a theorem of Herbrand, which is a sharpened version of Kummer's famous criterion. Explicitly, if B_{2k} denotes the 2k-th Bernoulli number $(B_2 = \frac{1}{6}, B_3 = 0, B_4 = \frac{1}{30}, \dots)$ and $2 \le 2k < N - 1$, write $j = 1 - 2k \mod N - 1$. Then:

THEOREM (Herbrand-Kummer): <u>If</u> B_{2k} <u>is a</u> N - <u>adic unit, then there are no nontrivial everywhere unramified</u> χ^j - <u>extensions of</u> $\mathbb{Q}(\zeta_N)$.

In particular, since $B_2 = 1/6$, Axiom 3 follows for $K = \mathbb{Q}$.

<u>Further remarks:</u> Ribet has recently shown [20b] that, under the same conditions as above, B_{2k} is an N - adic unit <u>if and</u> only if there are no nontrivial everywhere unramified χ^j - extensions, of $\mathbb{Q}(\zeta_N)$.

Gilles Robert [21] has developed machinery for the study of unramified χ^j - extensions of $K(\zeta_N)$ where K is a quadratic imaginary field. In particular he has a generalization of the Kummer criterion to this case, where the Bernoulli numbers are replaced by certain determinants of certain "Hurwitz numbers".

However, the (conjectured) analogue of Herbrand's "sharpening" is not yet known. [1)]

We may now proceed with the proof of the above proposition. For the remainder of the proof we shall let $E_{/K}$ denote an elliptic curve with a K - rational torsion point of order N (equivalently: a Galois sub-module isomorphic to the constant $\mathrm{Gal}(\overline{K}/K)$ - module $\mathbb{Z}/N$) and we shall study the properties of such a curve E, supposing that (K,N) satisfies our axioms. In the end we shall conclude that E cannot exist.

The e_N - pairing provides a self-duality of $E[N]$ into μ_N, and therefore our K - rational point of order N gives us an exact sequence of $\mathrm{Gal}(\overline{K}/K)$ modules:

$$(*) \qquad 0 \longrightarrow \mathbb{Z}/N \longrightarrow E[N] \longrightarrow \mu_N \longrightarrow 0$$

Choosing a $\mathbb{Z}/N$ - basis of $E[N](\overline{K})$ compatible with this exact sequence (i.e. such that the first member is a nontrivial K - rational point) enables us to view the 2 - dimensional $\mathrm{Gal}(\overline{K}/K)$ - representation over $\mathbb{Z}/N$ (the action of $\mathrm{Gal}(\overline{K}/K)$ on $E[N](\overline{K})$) as a representation $\rho: \mathrm{Gal}(\overline{K}/K) \longrightarrow GL_2\, \mathbb{Z}/N$ of the form $\begin{pmatrix} 1 & * \\ 0 & \chi \end{pmatrix}$, where χ is the standard character. Let L/K be the field extension generated by the N - division points of E (i.e. the 'splitting field' of the representation ρ). It is evident that L/K is Galois. The field L contains $K(\zeta_N)$, and from the exact sequence (*) one gets a natural injection

---

1) It might nevertheless be of interest to have lists of (K,N) where K is a quadratic imaginary field, and N is a rational prime, ≥ 5 remaining prime in K, such that N is a regular prime in K in the terminology of [21]. These (K,N) would indeed satisfy Axiom 3. ([21] Cor. 2).

$$(**) \qquad \mathrm{Gal}(L/K(\zeta_N)) \hookrightarrow \mathrm{Hom}(\mu_N(\overline{K}), \mathbb{Z}/N) \quad (= \mu_N^{-1}(\overline{K})) \quad .$$

To be sure, this shows that $\mathrm{Gal}(L/K(\zeta_N))$ is an abelian group killed by N. But a simple calculation shows, further, that the natural action of $\mathrm{Gal}(K(\zeta_N)/K)$ on $\mathrm{Gal}(L/K(\zeta_N))$ is by multiplication by the character χ^{-1}. As Serre pointed out, this calculation is particularly transparent if one views $\mathrm{Gal}(L/K(\zeta_N))$ as a subgroup of $\mu^{-1}(\overline{K})$ using (**) above.

Thus: $L/K(\zeta_N)$ <u>is a χ^{-1} - extension</u>.

We shall prove the following

MAIN LEMMA: (a) $L/K(\zeta_N)$ <u>is everywhere unramified.</u>

(b) E <u>is not an elliptic curve of complex multiplication.</u>

<u>Proof of the proposition, granted the main lemma:</u>

Since the χ^{-1} - extension $L/K(\zeta_N)$ is everywhere unramified, by Axiom 3, it is <u>trivial.</u> It follows that the exact sequence (*) splits, giving a $\mathrm{Gal}(\overline{K}/K)$ - isomorphism $E[N] = \mathbb{Z}/N \times \mu_N$. That is, we may view the $\mathrm{Gal}(\overline{K}/K)$ module $\mathbb{Z}/N \times \mu_N$ as contained in $E_{/K}$. We may pass to the quotient $E' = E/\mu_N$ which is, again, an elliptic curve over K, and the image of the sub-Galois module $\mathbb{Z}/N \subset E$ in E' is, again, a sub-Galois module isomorphic to $\mathbb{Z}/N$. Since $\mathbb{Z}/N \subset E'$ satisfies all the hypotheses that $\mathbb{Z}/N \subset E$ does, the main lemma is applicable to it also. Proceeding as above, we get a sequence of elliptic curves over K

$$\begin{array}{ccccccccc} E & \to & E' & \to & E'' & \to \cdots \to & E^{(j)} & \to \cdots \\ \cup\!\shortmid & & \cup\!\shortmid & & \cup\!\shortmid & & \cup\!\shortmid & \\ \mathbb{Z}/N & & \mathbb{Z}/N & & \mathbb{Z}/N & & \mathbb{Z}/N & \end{array}$$

each obtained from the next by an N - isogeny , and such that the original subgroup $\mathbb{Z}/N \subset E$ maps isomorphically into every $E^{(j)}$.

Since all the curves $E^{(j)}$ will have good reduction outside a fixed finite set of closed points of S = the spectrum of the ring of integers in K , it follows from Shafarevic's theorem ([22c] Ch. IV 1.4) that among the set of $E^{(j)}$'s there can be only a finite number of K - isomorphism classes of elliptic curves represented. Consequently, for some indices $j > j'$ we must have $E^{(j)} \cong E^{(j')}$. But $E^{(j)}$ maps to $E^{(j')}$ by a nonscalar isogeny. Therefore $E^{(j)}$, and hence E , is an elliptic curve of complex multiplication. But this contradicts part (b) of the main lemma.

<u>Remarks</u>: 1. The above argument, using part (a) of the main lemma, shows that E has a complex multiplication <u>defined over K</u>, which is impossible when $K = \mathbb{Q}$. So, in that case, one has a contradiction from part (a) alone.

2. Although Part (a) is an assertion which is 'local' for every prime of K , the essential step (2 below) in the proof of the main lemma is global.

<u>Step 1</u>: (<u>The Néron model of</u> $E_{/K}$)

Let S be the spectrum of the ring of integers in K , and $E_{/S}$ the Néron model of $E_{/K}$. By the universal property of Néron models the morphism

$\mathbb{Z}/N_{/K} \to E_{/K}$ extends to a morphism $\mathbb{Z}/N_{/S} \to E_{/S}$ which maps to the Zariski closure in $E_{/S}$ of $\mathbb{Z}/N_{/K} \subset E_{/K}$ (the 'group scheme extension' of $\mathbb{Z}/N_{/K}$ ([19] §2 ; [14a] Ch. 1 (c) .) This group scheme extension $G_{/S}$ is a (separated) quasi-finite group scheme over S whose generic fibre is $\mathbb{Z}/N$. Since, however, it admits a map from $\mathbb{Z}/N_{/S}$ which is an isomorphism on the generic fiber, it follows that $G_{/S}$ is a finite flat group scheme (of order N). Since, by Axiom 1, $N > d+1$, for each closed point $s \in S$, the absolute ramification index e_s (over Spec $\mathbb{Z}$) is $< N-1$, and consequently, by a theorem of Raynaud ([19] 3.3.6) $G_{/S} \cong \mathbb{Z}/N_{/S}$.

Therefore we shall identify $G_{/S}$ with $\mathbb{Z}/N_{/S}$, and we obtain, therefore, for each closed point $s \in S$ the subgroups $\mathbb{Z}/N_{/s} \subset E_{/s}$ in the Néron fibre over s (the 'specializations').

LEMMA 1: <u>$E_{/S}$ is semi-stable. That is, for each $s \in S$, $E_{/s}$ is either an elliptic curve, or its connected component $(E_{/s})^0$ is of multiplicative type.</u>

<u>Proof:</u> Suppose that $(E_{/s})^0$ is an additive group. Then the index of $(E_{/s})^0$ in $E_{/s}$ is ≤ 4 ([24 §6 Table p. 46). It follows that $\mathbb{Z}/N_{/s} \subset (E_{/s})^0$. Let $k(s)$ denote the residue field of s. Since the additive group over $k(s)$ is killed by multiplication by the characteristic of $k(s)$ (= "char s") it follows that char $s = N$. Now let K_s denote the completion of K at s, and note that there is a field extension K'_s/K_s whose relative ramification index is ≤ 6, and such that $E_{/K'_s}$ possess a semi-stable Néron model $\mathcal{E}/_{\mathcal{O}'_s}$ where $\mathcal{O}'_s$ is the ring of integers in

K'_s [23]. [1)] If $E_{/\mathfrak{O}'_s}$ denotes the pullback of $E_{/S}$ to $\mathfrak{O}'_s$, we have a morphism

$$E_{/\mathfrak{O}'_s} \xrightarrow{\varphi} \mathcal{E}_{/\mathfrak{O}'_s}$$

which is an isomorphism on generic fibres, using the Universal Néron Property of $\mathcal{E}_{/\mathfrak{O}'_s}$. The mapping φ is zero on the connected component of the special fibre of $E_{/\mathfrak{O}'_s}$ since there are no non-zero morphisms from an additive to a multiplicative type group over a field. Consequently, the mapping φ restricted to the special fibre of $\mathbf{Z}/N_{/\mathfrak{O}'_s}$ is zero. As in the discussion before the present lemma, one sees that if $\mathcal{G}_{/\mathfrak{O}'_s}$ is the 'group scheme extension' in $\mathcal{E}_{/\mathfrak{O}'_s}$ of $\mathbf{Z}/N_{/K'_s}$ then there is a morphism from $\mathbf{Z}/N_{/\mathfrak{O}'_s}$ to $\mathcal{G}_{/\mathfrak{O}'_s}$ which is an isomorphism on generic fibres, and which is zero on special fibres.

Using Raynaud's Cor. 3.3.6 [19], again, one sees that this is impossible, since the absolute ramification index of K'_s is $\leq 6d$ and $N - 1 > 6d$ by Axiom 1.

LEMMA 2: <u>If</u> $s \in S$ <u>is a point of characteristic</u> 2 <u>or</u> 3, <u>then</u> E <u>has bad (hence multiplicative) reduction over</u> s, <u>and</u> $\mathbf{Z}/N_{/s} \not\subset (E_s)^o$. (<u>Recall that</u> ^{o} <u>denotes connected component</u>).

---

1) <u>Proof:</u> apply §2 Corollary 3 of [23] with $m = 3$ <u>and</u> 4, noting that $N = \text{char } s$ is different from 2 and 3.

Proof: Let $d = [K : \mathbb{Q}]$. If s has characteristic ℓ then the cardinality of $k(s)$ is $\leq \ell^d$. If E_s has good reduction at s, it has at most $1 + \ell^d + 2 \cdot \ell^{d/2}$ points by the "Riemann hypothesis". Since $\mathbb{Z}/N \subset E_s$ this contradicts Axiom 1 if $\ell = 2$ or 3. Thus E has multiplicative type reduction at the point s. Then over the quadratic extension $\widetilde{k(s)}$ of $k(s)$, we have an isomorphism $(E_{/\widetilde{k(s)}})^o \cong \mathbb{G}_{m/\widetilde{k(s)}}$ ([22c] IV A. 1. 1) and therefore N must divide the cardinality of $\widetilde{k(s)}$. If $\mathrm{card}(\widetilde{k(s)}) = \ell^{2r}$ with $r \leq d$, then N divides $\ell^{2r} - 1 = (\ell^r - 1)(\ell^r + 1)$ which again violates Axiom 1, since N is prime.

Remark: We have established part (b) of the main lemma, since if E were a complex multiplication elliptic curve, its Néron model could not have multiplicative type reduction at any point $s \in S$.

Step 2: (The Global Step).

LEMMA 3: If $s \in S$ is any point of bad reduction for $E_{/S}$ then the 'specialization' $\mathbb{Z}/N_{/s}$ is not contained in the connected component of the identity $(E_{/s})^o$.

Proof: Let s_o be a point of bad reduction such that $\mathbb{Z}/N_{/s_o} \subset (E_{/s_o})^o$. By Lemma 1, $\mathrm{char}\, s_o \neq N$, and by Lemma 2, $\mathrm{char}\, s_o \neq 2, 3$. Let $T = \mathrm{Spec}\, \mathcal{O}[1/N]$ where $\mathcal{O}$ is the ring of integers in K, and let $X_0(N)_{/T}$ be the modular curve over the indicated base. It is a smooth scheme over T. Let $\underline{0}_{/T}$, $\underline{\infty}_{/T}$ denote the cuspidal sections of $X_0(N)_{/T}$ and let x denote the T-valued point of $X_0(N)$ determined by our couple $(\mathbb{Z}/N_{/T} \subset E_{/T})$. It is illuminating to draw a scheme-theoretic diagram:

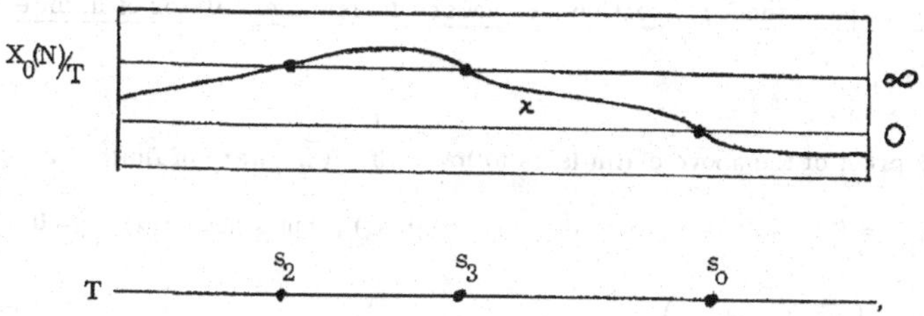

Here s_2 is any point of T of characteristic 2 and s_3 is any point of characteristic 3. Such points exist. We are justified in drawing the intersections $x_{/s_2} = \underline{\infty}_{/s_2}$ and $x_{/s_3} = \underline{\infty}_{/s_3}$ because, by ([4] VI §5) the modular interpretation of $\underline{\infty}_{/s}$ is the 'generalized elliptic curve' $(\mathbb{Z}/N \subset \mathbb{G}_m \times \mathbb{Z}/N)$ (i.e. the cyclic subgroup of order N which gives the $\Gamma_0(N)$ - structure is not contained in the connected component of the identity) while the interpretation of $\underline{0}_{/s}$ is the 'generalized elliptic curve' $(\mu_N \subset \mathbb{G}_m \times \mathbb{Z}/N)$ (i.e. the cyclic subgroup of order N which give the $\Gamma_0(N)$ - structure is contained in the connected component containing the identity).

Let us consider, now, any abelian variety quotient $B_{/\mathbb{Q}}$ of the jacobian J of $X_0(N)_{/\mathbb{Q}}$. We assume therefore the existence of a homomorphism $J \to B$ defined over $\mathbb{Q}$. Since J has 'good reduction' over Spec $\mathbb{Z}[1/N]$, by the Criterion of Néron-Ogg-Shafarevic (cf. [23] §1), the Néron model of B over Spec $\mathbb{Z}[1/N]$ is an abelian scheme, and consequently so is its pullback $B_{/T}$. We have a morphism $X_0(N)_{/T} \xrightarrow{f} B_{/T}$ which sends $\underline{\infty}_{/T}$ to the zero-section. This morphism is simply the composition of the natural morphisms $X_0(N)_{/T} \xrightarrow{\alpha} J_{/T} \longrightarrow B_{/T}$ where α sends a section z to the divisor class of $z - \underline{\infty}$.

Claim: The image of the T - section x under f is either 0 or of infinite order in $B(T)$.

The proof of the above claim is as follows: If $f(x)$ were of finite order m, since $\gamma(x)_{/s_2} = 0$, m is a power of 2 (= char s_2) ; but since $\gamma(x)_{/s_3} = 0$, m is a power of 3 (= char s_3).

We are now ready to invoke Axiom 3. Applying the above Claim to the abelian variety $B = A$ of Axiom 3, we deduce that $f(x) = 0$. Consequently $f(\underline{0}_{/s_o}) = 0$ (in $B_{/s_o}$). But since f is defined over the base $S' = \operatorname{Spec} \mathbb{Z}[1/N]$ and $\underline{0} - \underline{\infty}$ is an S' - section of finite order $\neq 0$, one obtains by Oort-Tate [18] or by Raynaud ([19] 3.3.6) that $f(\underline{0}_{/s'}) \neq 0$ for any point $s' \in S'$ of characteristic $\neq 2$. Taking s' to be the image of s_o in S', we arrive at a contradiction, and we conclude the assertion of Lemma 3.

Step 3: ($L/K(\zeta_N)$ is unramified)

Proof: We shall prove the above assertion for all closed points $s \in S$:

(a) If E has good reduction at s, and char $s \neq N$ then $E[N]$ is an étale finite flat group scheme in a Zariski open set about s. In particular, L/K is unramified 'above' the point $s \in S$.

(b) If E has good reduction at s, and char $s = N$ then $E[N]$ is a finite flat group scheme over S_s, the completion of S at the point s. Applying the connected component of the identity functor (denoted ^{o}) to the exact sequence of finite flat group schemes over S_s : $0 \to \mathbb{Z}/N \to E[N] \to \mu_N \to 0$, one sees that

$E[N]^o = \mu_N$, giving us a canonical splitting: $E[N] = \mathbb{Z}/N \times \mu_N$ which shows, again, that $L/K(\zeta_N)$ is unramified 'above' the point $s \in S$.

(c) If E has bad (hence multiplicative) reduction at s , we shall work, as in (b) , over the base S_s . The quasi-finite group scheme $E[N]_{/S_s}$ (= kernel of multiplication by N in the group scheme $E_{/S_s}$) fits into a short exact sequence of quasi-finite group schemes over S_s : $0 \to \mathbb{Z}/N \to E[N] \to G \to 0$ where the generic fibre of G is isomorphic to μ_N . The point of Lemma 3 in Step 2 is to insure that the special fibre of G is non-zero. Explicitly, since $\mathbb{Z}/N_{/s} \not\subset (E_{/s})^o$, we have that the kernel of N in the multiplicative group $(E_{/s})^o$ maps injectively to $G_{/s}$. It follows that G , and hence $E[N]$ is a finite flat group scheme over S_s . If char $s \neq N$ then $E[N]$ is an étale finite flat group scheme, and one concludes as in (a) above. If char $s = N$, then let us note that $G = \mu_N$ over S_s . (Here are two possible arguments for this: By Axiom 1 , the absolute ramification index of K_s is $< N - 1$, and therefore a finite flat group scheme over S_s of order N is determined by its generic fibre ([19] 3.3.6) . Or, one can show directly that $E[N]$, being a finite flat group scheme must be self (Cartier) dual, using an autoduality formula for Néron models and 'Néron-connected' models. [15] Ch. I 5.1) .

We thus have a short exact sequence of finite flat group schemes as in (b) above, and we conclude the argument similarly.

§4. Eisenstein quotients.

Our presentation of the proof of Theorem 1 would be incomplete without some account of the proof that the pair $(\mathbb{Q}, N)$ satisfies Axiom 2 when $N = 11$ or $N \geq 17$ (i.e. when the genus of $X_0(N)$ is > 0). Two different proofs of this are given in [14a]. (Let us call these the easy and hard proofs. The 'easy' proof is also sketched in [16]. It is given in [14a] in Ch. III §3). Both proofs rely on an argument of 'geometric descent' using the 'Eisenstein ideal' in the Hecke algebra of endomorphisms of J the jacobian of $X_0(N)$. The easy proof is an 'infinite descent' (a more appropriate title would be: an 'indefinite descent') which uses surprisingly little information concerning the Eisenstein ideal. The hard proof uses the detailed study of the Eisenstein ideal given in Ch. II of [14a]. It is a 'first descent' and yields more precise information concerning the Shafarevic-Tate group.

Rather than repeat either of these proofs here, we shall adapt the hard proof so as to make it yield information in the case where K is a quadratic imaginary field. In particular, we shall prove that (K, N) satisfies Axiom 2 (ii) (i.e. the Eisenstein quotient of J has a nontrivial factor with finite Mordell-Weil group over K) provided K is a quadratic imaginary field in which N does not split, and $N \geq N(K)$ where $N(K)$ is an explicit constant, dependent upon K (Cor. 2 below). The easy proof would not suffice for this application. The reader should easily be able to reconstitute the hard proof for $K = \mathbb{Q}$ from the facts concerning the Eisenstein ideal collected below, and the proof given.

We shall try to introduce the reader to the relevant parts of the theory of the

Eisenstein ideal by presenting the needed definitions interspersed with results quoted from [14a] (collected in the facts numbered 1 - 6 below). Having accumulated what we need, it will be a relatively simple matter to 'perform the required descent'.

To begin, N will denote a fixed prime number such that genus $(X_0(N)) > 0$ (i.e. $N = 11$ or $N \geq 17$). Let J be the jacobian of $X_0(N)$ over $\mathbb{Q}$, and $J_{/\mathbb{Z}}$ its Néron model over $\mathbb{Z}$.

$\mathbb{T}$: The Hecke algebra is the subalgebra of the endomorphism ring $\mathrm{End}_{\mathbb{C}} J$ generated by the Hecke operators T_ℓ (ℓ running through all rational prime numbers $\neq N$) and by w (the canonical involution induced from $z \mapsto -1/Nz$ on the upper half plane $\mathbb{H}$).

Fact 1: $\mathbb{T} = \mathrm{End}_{\mathbb{C}} J$ ([14a] Ch. II 9.5).

I: The Eisenstein ideal is the ideal in $\mathbb{T}$ generated by the elements

$$\eta_\ell = 1 + \ell - T_\ell \qquad (\ell \neq N)$$

and by $1 + w$.

Fact 2: By ([14a] Ch. II 9.7) one has $\mathbb{T}/I = \mathbb{Z}/n\mathbb{Z}$ where $n = \text{numerator}\left(\frac{N-1}{12}\right)$. Let us reserve the letter p to denote a rational prime number dividing n. The prime ideals of $\mathbb{T}$ which are in the support of I (called the Eisenstein primes) are in one-one correspondence with the prime divisors p of n.

Let P be the (Eisenstein) prime generated by I and p. Then $\mathbb{T}/P = \mathbb{Z}/p$.

Fact 3: The Eisenstein ideal is locally principal in $\mathbb{T}$ ([14a] Ch. II 18.10). Explicitly we have the following criterion which furnishes us amply with local generators of I at an Eisenstein prime $P = (I, p)$:

Let (p, ℓ) be a pair of rational primes $\neq (2,2)$ such that p divides n. Then the element $\eta_\ell = 1 + \ell - T_\ell$ is a generator of the ideal I locally at $P = (I, p)$ if and only if:

(i) ℓ is not a p-th power mod N

(ii) $\frac{\ell - 1}{2} \not\equiv 0 \mod p$.

In the exceptional case $(p, \ell) = (2,2)$ we have that η_2 is a local generator of I at $P = (I, 2)$ if and only if 2 is not a quartic residue modulo N.

C: the cuspidal subgroup. If c is the class of the divisor of degree zero $0 - \infty$ in $J(\mathbb{Q})$, then ([17a], [14a] Ch. II 11.1) n = order (c). The cuspidal subgroup C is the subgroup of $J(\mathbb{Q})$ generated by c. We use the notation $C_{/\mathbb{Z}}$ to indicate the finite flat subgroup scheme over $\mathbb{Z}$ generated by C in $J_{/\mathbb{Z}}$. By ([14a] Ch. II 11.1) C is annihilated by the Eisenstein ideal.

$J[I]_{/\mathbb{Q}}$: the kernel of the ideal I in the jacobian $J_{/\mathbb{Q}}$. This is, by definition, the intersection of the kernels in $J_{/\mathbb{Q}}$ of all (or of a generating system of) elements in I.

Fact 4: By ([14a] Ch. II 16.4 and 17.9) $J[I]_{/\mathbb{Q}}$ is of order n^2. By ([14a] Ch. II 1.7) there is a Galois submodule $\Sigma \subset J[I]_{/\mathbb{Q}}$ (called the Shimura subgroup) such that Σ is isomorphic (as $\mathrm{Gal}(\overline{\mathbb{Q}}/\mathbb{Q})$ - module) to μ_n. If n is odd (i.e. $N \not\equiv 1 \mod 8$) then $J[I] = C \oplus \Sigma$. If, however, n is even, then $C \cap \Sigma$ is a sub-Galois module of order 2 and thus $C + \Sigma$ is of index two in $J[I]$. Only in certain cases (i.e. $N \equiv 9 \mod 16$) have we given an explicit construction of the 'remaining piece' in $J[I]$.

Fact 5: (The fibre of Néron in characteristic N)

The 'bad' Néron fiber $J_{/\mathbb{F}_N}$ has the following structure:

$$J_{/\mathbb{F}_N} = (J_{/\mathbb{F}_N})^0 \times \overline{C}$$

where $\overline{C}$ is a cyclic group of order n, which may be viewed as the specialization to $\mathbb{F}_N$ of the cuspidal group C, and $(J_{/\overline{\mathbb{F}}_N})^0$ is a multiplication type group. ([14a] Appendix).

Fact 6: (Quotients of J) An abelian variety quotient of J, $J_{/\mathbb{Q}} \xrightarrow{f} B_{/\mathbb{Q}}$ will be called an optimal quotient if the kernel of f is an abelian subvariety of J (i.e. if it is connected). Clearly every quotient is isogenous to a unique optimal quotient. The $\mathbb{Q}$-simple quotients of J are $\mathbb{C}$-simple [20a].

The Hecke algebra $\mathbb{T}$ has the property that $\mathbb{T} \otimes \mathbb{Q} = \prod_j k_j$ (***) where k_j are (totally real) algebraic number fields. One has the following natural one-one correspondences:

$$\left\{\begin{matrix}\text{simple optimal}\\ \text{quotients of } J\end{matrix}\right\} \longleftrightarrow \left\{\begin{matrix}\text{factors } k_j\\ \text{in } (***)\end{matrix}\right\} = \left\{\begin{matrix}\text{irreducible}\\ \text{components in } \operatorname{Spec} \mathbb{T}\end{matrix}\right\}$$

$$= \left\{\begin{matrix}\text{minimal prime ideals}\\ \text{of } \mathbb{T}\end{matrix}\right\}$$

If $\mathfrak{a}$ is any ideal in $\mathbb{T}$ let $\gamma_{\mathfrak{a}} \subset \mathbb{T}$ be the ideal $\gamma = \bigcap_{r=1}^{\infty} \mathfrak{a}^r$ and $J^{(\mathfrak{a})}/_{\mathbb{Q}}$ the optimal quotient of J obtained by passing to the quotient of J by the abelian subvariety $\gamma_{\mathfrak{a}} \circ J \subset J$. Geometrically, we may view $J^{(\mathfrak{a})}$ as that optimal quotient of J, which under the one-one correspondence above corresponds to the set of all those irreducible components of Spec $\mathbb{T}$ which meet the support of $\mathfrak{a}$.

$\tilde{J}$: The Eisenstein quotient; it is the optimal quotient $J^{(I)}$ where I is the Eisenstein ideal.

$\tilde{J}^{(p)}$: The p - Eisenstein quotient, $J^{(P)}$, where $P = (I, p)$.

One has that the Eisenstein quotient is that optimal quotient of J comprising all the simple quotients of $\tilde{J}^{(p)}$, $p | n$.

For a detailed study of these quotients and numerical data for $N < 250$ see [14a].

ℓ_P, ℓ_p : If W is a finite $\mathbb{T}$ - module, $\ell_P(W)$ is the P - length of W. If W is a finite abelian group then $\ell_p(W)$ denotes its p - length (i.e. $\log_p$ of the order of the p - Sylow subgroup of W).

$H^i(S, \mathfrak{F})$: will denote cohomology for the fppf site ([26] exp. IV 6.3) over a scheme S, where $\mathfrak{F}$ is an abelian fppf - sheaf. The reader will note that the only fppf sheaves we use explicitly are flat group schemes over S (although Φ, which occurs below, is an étale nonseparated group scheme). Moreover, the only dimensions i we consider are: $H^0(S,\mathbb{F})$ = group of S - valued sections of $\mathbb{F}$, and $H^1(S,\mathbb{F})$.

Conventions: Fix p a prime divisor of n, and $\eta = \eta_\ell$ a local generator of I at $P = (I, p)$ (See fact 3). One sees by an elementary argument that η is an isogeny of J (cf. [14a] Ch. II. proof of 16. 10).

Let $\Delta \subset \operatorname{Spec} \mathbb{T}$ be the closed subscheme (the finite set of closed points) whcih is the complement of the point P in the support of the ideal (η). We shall work consistently modulo Δ. That is, we shall ignore finite $\mathbb{T}$ - modules supported on Δ.

Let K be a quadratic imaginary field and S the spectrum of the ring of integers in K.

Consider the exact sequence of abelian fppf sheaves over S:

(the "descent sequence"): $0 \longrightarrow J[\eta] \longrightarrow J \xrightarrow{\eta} J \longrightarrow \Phi \longrightarrow 0$.

Here $J[\eta]$ is the kernel of η in $J_{/S}$. It is a quasi-finite (separated) flat group scheme. The cokernel of η, Φ, is a 'skyscraper sheaf' concentrated at the points s of S of characteristic N. Its stalk at any such point is isomorphic to $\overline{C}$ (fact 5).

Now let p^α denote the maximal power of p dividing n. Thus, $\ell_p(C) = \ell_p(C) = \alpha$. If $p \neq 2$, $J[\eta] \cong \mathbb{Z}/p^\alpha \oplus \mu_{p^\alpha}$ modulo Δ (fact 4).

Let h(K) denote the class number of K and $\beta = \ell_p(h(K))$. We make the further hypothesis that $\alpha \geq \beta$ which will be strengthened later.

The Mordell-Weil group of J over K is the finitely generated group $H^0(S,J) = J(S) = J(K)$, which we view as a $\mathbb{T}$ - module (indeed: as a coherent sheaf over Spec $\mathbb{T}$). Let $M = J(K)/\underline{\text{torsion}}$.

Set $\nu = 0$ if N does not split in K, (i.e. ramifies or stays prime), and $\nu = 1$ if N splits in K.

The descent estimate:

$$\ell_p(M/\eta \cdot M) \leq 2\beta + \nu \cdot \alpha \qquad \text{if } p \neq 2, \text{ and}$$

$$\ell_p(M/\eta \cdot M) \leq 2\beta + \nu \cdot \alpha + (1 + g + \nu) \qquad \text{if } p = 2,$$

where g is the 2-length of the subgroup of points of order two in the ideal class group of K.

Proof: We indicate the proof in some detail when $p \neq 2$. When $p = 2$, we 'lose' the quantity $(1+g+\nu)$ in our estimate since we lack a complete description of the P - primary component of $J[I]$ and possess a description only 'up to a group of order two'.[1)]

Suppose, then, $p \neq 2$.

Note that $\ell_p(W)$ depends only on W modulo Δ. The estimate is established by first obtaining a bound for the P - length of $J(K)/\eta \cdot J(K)$, by estimating the P - lengths of terms occuring in the long (fppf) - cohomological exacts sequences arising from the "descent sequence". For this calculation one

---

1) To complete the argument for $p = 2$ when $K = \mathbb{Q}(\sqrt{-1})$ one must make use of the explicit Galois module structure of the points of order 2 in $J[I]$ ([14a] Ch. II §12).

must know that:

(a) $\ell_p(H^1(S,\mathbb{Z}/p^{\alpha})) = \beta$,

for (since $\alpha \geq \beta$) $H^1(S,\mathbb{Z}/p^{\alpha})$ is isomorphic to the dual of the p - primary component of the Hilbert Class Field of K.

(b) $\ell_p(H^1(S,\mu_{p^\alpha})) = \beta$ if $(p,K) \neq (3, \mathbb{Q}(\sqrt{-3}))$

$= \beta + 1$ if $(p,K) = (3, \mathbb{Q}(\sqrt{-3}))$.

Proof: By Kummer theory for μ_{p^α}, we have the short exact sequence:

$$0 \longrightarrow S^*/S^{*p^\alpha} \longrightarrow H^1(S,\mu_{p^\alpha}) \longrightarrow H^1(S,\mathbb{G}_m)[p^\alpha] \longrightarrow 0$$

where S^* denotes $\mathbb{G}_m(S)$ = Global units in K, and $[p^\alpha]$ means, as usual, the kernel of multiplication by p^α.

If we recall that $H^1(S,\mathbb{G}_m)$ is the ideal class group of K, and, again, that $\alpha \geq \beta$, we obtain (b).

(c) $H^0(S,\Phi) = \overline{C}$ if N does not split in K.

$= \overline{C} \oplus \overline{C}$ if N does split in K.

$\ell_p(H^0(S,\Phi)) = (1+\nu)\cdot\alpha$.

From the "descent sequence" and (a), (b), and (c) one may deduce that $\ell_p(J(K)/\eta\cdot J(K))$ is $\leq (1+\nu)\alpha + 2\beta + \epsilon$ where $\epsilon = 1$ if $(p,K) = (3, \mathbb{Q}(\sqrt{-3}))$ and $\epsilon = 0$ otherwise. By applying the snake-lemma to the endomorphism η operating on the short exact sequence $0 \to \underline{\text{torsion}} \to J(K) \to M \to 0$, one obtains the "descent estimate" above.

If $\mathcal{P} \subset \mathbb{T}$ is a prime ideal, let the subscript $(\mathcal{P})$ denote localization at $\mathcal{P}$.

PROPOSITION: Suppose

$$\alpha > 2\beta \quad \text{if } p \neq 2$$

$$\alpha > 2\beta + (1 + g + \nu) \quad \text{if } p = 2 .$$

Then there is a minimal prime ideal $\mathcal{P} \subset P \subset \mathbb{T}$ such that the $\mathbb{T}_{(\mathcal{P})}$ - rank of $M_{(\mathcal{P})}$ is 0 if N does not split in K, and is ≤ 1 if N splits in K.

Proof: Comparing our hypotheses wtih the descent estimates we see that we have $(1+\nu)\alpha = (1+\nu)\,\ell_p(\mathbb{T}/\eta \cdot \mathbb{T}) > \ell_p(M/\eta \cdot M)$. Consequently,

$$(1+\nu) \cdot \ell_p(\mathbb{T}_{(P)}/\eta \cdot \mathbb{T}_{(P)}) > \ell_p(M_{(P)}/\eta \cdot M_{(P)}) .$$

Claim: $M_{(P)}$ does not contain a free $\mathbb{T}_{(P)}$ - module of rank $1+\nu$.

Proof: If $F \subset M_{(P)}$ is such a $\mathbb{T}_{(P)}$ - module, and Q is the quotient of $M_{(P)}$ by F, apply the snake-lemma to the endomorphism η operating on the exact sequence $0 \to F \to M_{(P)} \to Q \to 0$ and one quickly deduces a contradiction to the inequality displayed above.

Our proposition then follows from the claim, for if R is a commutative noetherian local subring of $R \otimes \mathbb{Q}$ = a product of fields, and if W is an R - module of finite type, then W contains a free R - module of rank r if and only if $W_{(\mathfrak{P})}$ is free of rank $\geq r$ over $R_{(\mathfrak{P})}$ for every minimal prime $\mathfrak{P}$ in R.

COROLLARY 1: If the inequalities of the previous proposition hold, and if, further, N does not split in K, then there is an (optimal) abelian variety quotient $\tilde{J}^{(p)} \to A$ defined over $\mathbb{Q}$, such that $A(K)$ is finite.

Proof: This follows directly from the proposition (cf. [14a] Ch. III 3.5).

Remarks: 1. If $p \neq 2$, and in the frequently encountered case $\beta = 0$, the above argument can be made to show that $\tilde{J}^{(p)}$ itself has a finite Mordell-Weil group over K. One has no reason to believe that this will continue to be true when $\beta > 0$.

Nevertheless it seems difficult to get examples where $\tilde{J}^{(p)}$ is not simple. The only example of this when $N < 250$ is for $p = 2$, $N = 113$ (See the table in the introduction of [14a]). If one admits certain standard conjectures (of Weil, and Hardy-Littlewood. [14a] Ch. III §7) one sees, however, that $\tilde{J}^{(2)}$ is not simple for an infinite number of values of N.

2. It seems likely that, if N does split in K, the $\mathbb{T}_{(\mathfrak{P})}$ rank of $M_{(\mathfrak{P})}$ is ≥ 1 for every minimal prime $\mathfrak{P} \subset P$. [1)]

---

1) If $x \in X_0(N)$ is represented by an elliptic curve with complex multiplication by the ring of integers in K, with N-isogeny given by one of its complex multiplications, there is some evidence to support the hope that the trace to K of the class $x - \infty$ in J generates a $\mathbb{T}_{(\mathfrak{P})}$-vectorspace of dimension one in $M_{(\mathfrak{P})}$ for every minimal prime $\mathfrak{P} \subseteq \mathbb{T}$.

As a consequence of the proposition one would then have the existence of a minimal prime $\wp$ such that the $\mathbb{T}_{(\wp)}$ - rank of $M_{(\wp)}$ is precisely 1.

3. For a fixed quadratic imaginary number field K, the inequalities required by the proposition will hold for some prime divisor p of $n = \mathrm{num}(\frac{N-1}{12})$ for all but a finite number of values of N. (e.g. $N > 48 \cdot h(K)^3 + 1$ will certainly insure the existence of such a p.)

COROLLARY 2: If $N > 48 \cdot h(K)^3 + 1$, and N does not split in the quadratic imaginary field K, then $X_0(N)(K)$ is finite.

Proof: In this case Cor. 1 applies, giving a nonconstant map $X_0(N) \xrightarrow{f} A$ (defined over $\mathbb{Q}$) where A is an abelian variety such that $A(K)$ is finite. Since $X_0(N)$ is of dimension one, the fibers of the mapping f are finite. Therefore $X_0(N)(K)$ is also finite.

4. (Examples of isogenies over quadratic imaginary fields.) Consider only prime numbers N such that genus $X_0(N) > 1$. Let $X^+ = X_0(N)^+$ denote the quotient of $X = X_0(N)$ by the canonical involution w. Since N is prime, it is known that the real locus $X(\mathbb{R})$ consists in a single circle, and if $X^+(\mathbb{R})^o$ is the connected components in $X^+(\mathbb{R})$ containing the image of the cusps, then the natural projection sends $X(\mathbb{R})$ to a proper arc in $X^+(\mathbb{R})^o$ (since w has a fixed point in $X(\mathbb{R})$). Call the complement of this image the imaginary arc in $X^+(\mathbb{R})^o$. Any $\mathbb{Q}$ - rational point of X^+ in this imaginary arc will provide (by passing to the inverse image in X) an N - isogeny, rational over some quadratic

imaginary field. When are there an infinite number of $\mathbb{Q}$ - rational points of X^+ lying in the imaginary arc? This will certainly be the case when X^+ is of genus 0 (N = 23, 29, 31, 41, 47, 59, and 71). This will also be the case when X^+ is of genus 1 (N = 37, 43, 53, 61, 79, 83, 89, 101, and 131). For, if J^+ is the jacobian of X^+, it is proved in [14a] (introduction. Theorem 3) that if the genus of X^+ is > 0, then the Mordell-Weil group of J^+ is a free abelian group of positive rank. Thus, in particular, when X^+ is an elliptic curve, its Mordell-Weil group is infinite [1)] and therefore its intersection with the circle group $X^+(\mathbb{R})^o$ (which is at most of index 2 in $X^+(\mathbb{R})$) must likewise be infinite, hence dense.

It would be interesting to obtain N - isogenies (prime N) over quadratic imaginary fields which do not arise from the above process nor from complex multiplication. In this connection one might mention that there are four values of N known (N = 389, 419, 479 and 491) such that $X_0(N)$ has only a finite number of cubic points. That is, the totality of rational points of $X_0(N)$ in all cubic fields is a finite set ([14a] Ch. III 4.6, using data provided by Atkin on New Year's eve 1975). Does this persist for larger values of N ?

1) Brumer and Kramer have shown it to be infinite cyclic.

BIBLIOGRAPHY

1. Berkovic, V.: On rational points on the jacobians of modular curves [in Russian]. To appear.

2. Brylinski, J.-L.: Torsion des courbes elliptiques (d'après Demjanenko). D.E.A. de Mathématique Pure presented at the Faculté des Sciences de Paris-Sud (1973).

3. Cassels, J. W. S.: Diophantine equations with special reference to elliptic curves. J. London Math. Soc. 41 (193-291) (1966).

4. Deligne, P., Rapoport, M.: Schémas de modules des courbes elliptiques. Vol. II of the Proceedings of the International Summer School on Modular Functions, Antwerp (1972). Lecture Notes in Mathematics 349. Berlin-Heidelberg-New York: Springer 1973.

5. Demjanenko, V. A.: Torsion of elliptic curves [in Russian], Izv. Akad. Nauk. CCCP, 35, 280-307 (1971) [MR 44, 2755].

6. Dörrie, H.: 100 great problems of elementary mathematics; their history and solution. Dover, New York 1965.

7. Griffiths, P.: Variations on a theme of Abel. Inventiones Math. 35 321-390 (1976).

8. Hellegouarch, Y.: Courbes elliptiques et équation de Fermat. Thèse d'Etat. Faculté des Sciences de Besançon (1972). See also the series of notes in the Comptes-Rendus de l'Académie des Sciences de Paris. 260 5989-5992, 6256-6258 (1965); 273 540-543, 1194-1196 (1971).

9. Herbrand, J.: Sur les classes des corps circulaires. Journal de Math. Pures et Appliquées. 9e série II, 417-441 (1932).

10. Kubert, D.: Universal bounds on torsion of elliptic curves. Proc. London Math. Soc. (3) 33 193-237 (1976).

11. Kubert, D., Lang, S.: Units in the modular function field. I, II, III Math. Ann. 218, 67-96, 175-189, 273-285 (1975).

12. Lang, S.: Elliptic Functions. Addison Wesley, Reading 1974.

13. Manin, Y.: A uniform bound for p - torsion in elliptic curves [in Russian]. Izv. Akad. Nauk. CCCP, 33 459-465 (1969).

14a. Mazur, B.: Modular curves and the Eisenstein Ideal. To appear: Publ. Math. I.H.E.S.

14b. Mazur, B.: p - adic analytic number theory of elliptic curves and abelian varieties over $\mathbb{Q}$. Proc. of International Congress of Mathematicians at Vancouver, 1974, vol. I, 369-377, Canadian Math. Soc. (1975).

15. Mazur, B., Messing, W.: Universal extensions and one dimensional crystalline cohomology. Lecture Notes in Mathematics. 370. Berlin-Heidelberg-New York: Springer 1974.

16. Mazur, B., Serre, J.-P.: Points rationnels des courbes modulaires $X_0(N)$. Séminaire Bourbaki no. 469. Lecture Notes in Mathematics. 514 Berlin-Heidelberg-New York: Springer 1976.

17a. Ogg, A.: Rational points on certain elliptic modular curves. Proc. Symp. Pure Math. 24 221-231 (1973) AMS, Providence.

17b. Ogg, A.: Diophantine equations and modular forms. Bull. AMS 81 14-27 (1975).

18. Oort, F., Tate, J.: Group schemes of prime order. Ann. Scient. Ec. Norm. Sup. série 4, 3, 1-21 (1970).

19. Raynaud, M.: Schémas en groupes de type $(p,\cdots,p)$. Bull. Soc. Math. France. 102 fasc. 3, 241-280 (1974).

20a. Ribet, K.: Endomorphisms of semi-stable abelian varieties over number fields. Ann. of Math. 101 no. 3. 555-562 (1975).

20b. Ribet, K.: A modular construction of unramified p - extension of $\mathbb{Q}(\mu_p)$. Inventiones Math. 34, 151-162 (1976).

21. Robert, G.: Nombres de Hurwitz et regularité des idéaux premiers d'un corps quadratique imaginaire. Séminaire Delange-Pisot-Poitou. Exposé given April 28, 1975.

22a. Serre, J.-P.: Propriétés galoisiennes des points d'ordre fini des courbes elliptiques. Inventiones math. 15, 259-331 (1972).

22b. Serre, J.-P.: p - torsion des courbes elliptiques (d'après Y. Manin) Séminaire Bourbaki 69/70 no. 380. Lecture Notes in Mathematics. 180. Berlin-Heidelberg-New York: Springer 1971.

22c. Serre, J.-P.: Abelian ℓ - adic representations and elliptic curves. Lectures at McGill University. New York-Amsterdam: W. A. Benjamin Inc., 1968.

23. Serre, J.-P., Tate, J.: Good reduction of abelian varieties. Ann. of Math. 88, 492-517 (1968).

24. Tate, J.: Algorithm for determining the Type of a Singular Fiber in an Elliptic Pencil. 33-52. Modular Functions of one variable IV. Proceedings of the International Summer School, Antwerp RUCA. Lecture Notes in Mathematics 476. Berlin-Heidelberg-New York: Springer 1975.

25. SGA 3: Schémas en groupes I. Lecture Notes in Mathematics. 151. Berlin-Heidelberg-New York: Springer 1970.

B. Mazur
Harvard University
Department of Mathematics
Science Center
One Oxford Street
Cambridge, Mass. 02138

COURBES MODULAIRES DE NIVEAU 11

par Gérard LIGOZAT

INTRODUCTION

On appelle groupe de congruence de niveau 11 tout sous-groupe de $SL_2(\mathbb{Z})$ qui contient le groupe $\Gamma(11)$ des matrices congrues à la matrice unité modulo 11.

Soient $X_{\text{dép}}(11)$, resp. $X_{\text{ndép}}(11)$, resp. $X_{\mathfrak{S}_4}(11)$ les courbes algébriques associées aux groupes de congruence dont l'image dans $PSL_2(\mathbb{F}_{11})$ est le normalisateur d'un groupe de Cartan déployé, resp. non déployé, resp. d'un groupe isomorphe au groupe $\mathfrak{A}_4$, groupe alterné de quatre éléments. Il existe une façon canonique de munir ces courbes d'une structure de courbe sur $\mathbb{Q}$. Une conséquence de ce travail est le résultat suivant :

(1) *La courbe $X_{\text{dép}}(11)$ possède un nombre fini de points rationnels sur $\mathbb{Q}$; les points rationnels sur $\mathbb{Q}$ de $X_{\text{ndép}}(11)$, resp. $X_{\mathfrak{S}_4}(11)$ forment un groupe isomorphe à $\mathbb{Z}$, resp. un groupe réduit à l'élément neutre*.

Pour démontrer ceci, on se ramène à considérer des courbes elliptiques dont on connait le groupe des points rationnels, grâce aux tables de [2]. Plus précisément, considérons par exemple la courbe $X_{\text{ndép}}(11)$. C'est une courbe elliptique, définie sur $\mathbb{Q}$, et qui admet une bonne réduction en dehors de $\{11\}$. D'après une conjecture de Weil, une telle courbe est un quotient sur $\mathbb{Q}$ de la jacobienne $J_o(121)$ de la courbe modulaire $X_o(121)$.

Ceci nous amène à étudier, dans un premier temps, les courbes elliptiques qui sont des quotients de $J_o(121)$. La liste de ces courbes est essentiellement connue, grâce à Swinnerton-Dyer, Vélu et aux calculs de Tingley (cf. [2] et [16]). Cependant, nous avons besoin de renseignements supplémentaires sur la structure du groupe engendré dans $J_o(121)$ par les pointes de $X_o(121)$. Pour déterminer cette structure, on utilise les propriétés des symboles modulaires ainsi que la description de la variété abélienne $J_o(121)$ en termes des périodes des formes paraboliques attachées à $\Gamma_o(121)$. C'est de cette façon que Manin démontre le fait que la différence de deux pointes est d'ordre fini dans la jacobienne. Comme G. Goldstein l'a fait remarquer à l'auteur, cette méthode permet en fait la détermination explicite de la classe d'homologie du chemin joignant deux pointes.

Nos calculs fournissent une démonstration directe, indépendante de la détermination d'équations explicites, du résultat suivant : les courbes figurant dans la liste de Swinnerton-Dyer et Vélu sont les courbes elliptiques quotients de $J_o(121)$; ils nous permettent de retrouver, et de préciser les propriétés de ces courbes.

En particulier, chacune des courbes considérées est caractérisée par des propriétés simples de ses points de torsion. Plus précisément :

(2) *Soit* E *une courbe elliptique définie sur* $\mathbb{Q}$, *quotient sur* $\mathbb{Q}$ *de* $J_o(121)$. *Alors on est dans l'un des deux cas suivants* :

(a) *la courbe* E, *ou la courbe tordue de* E *sur* $\mathbb{Q}(\sqrt{-11})$, *est isogène sur* $\mathbb{Q}$ *à* $X_o(11)$; *dans ce cas, les propriétés galoisiennes des points d'ordre* 25 *caractérisent* E ;

(b) *la courbe* E *contient un groupe d'ordre* 11 *rationnel sur* $\mathbb{Q}$; *la structure galoisienne de ce groupe caractérise* E .

Ceci fait, deux méthodes peuvent être utilisées pour montrer que $X_{ndép}(11)$ est effectivement l'une des courbes dont on vient de faire la liste, et pour déterminer de quelle courbe il s'agit.

- La première consiste à étudier le groupe engendré par les pointes de $X_{ndép}(11)$. Pour cela, on utilise la construction explicite de fonctions dont le diviseur est concentré aux pointes. On sait grâce à Kubert et Lang [5] que les formes de Klein sont des outils bien adaptés pour cette construction, en ce sens que toute fonction inversible et non nulle en dehors des pointes est essentiellement un produit de formes de Klein.

On montre ainsi que $X_{ndép}(11)$ contient un groupe d'ordre 11, rationnel sur $\mathbb{Q}$; de plus, on a une description de la structure galoisienne de ce groupe. Ceci suffit à montrer que $X_{ndép}(11)$ est l'une des courbes de la liste, et plus précisément que c'est l'une des courbes dont le groupe des points rationnels sur $\mathbb{Q}$ est isomorphe à $\mathbb{Z}$; d'où le résultat annoncé.

- Une seconde méthode consiste à adopter le point de vue des représentations de Hecke. Plus précisément, considérons la représentation de $PSL_2(\mathbb{F}_{11})$ réalisée par les formes paraboliques de poids 2 sur le groupe $\Gamma(11)$. On sait, grâce à Hecke, décomposer cette représentation en somme de ses composantes irréductibles. On constate que chacune de ces composantes correspond à une classe d'isogénie sur $\mathbb{Q}(\sqrt{-11})$ de courbes elliptiques quotients de $J_o(121)$. La détermination de $X_{ndép}(11)$ à $\mathbb{Q}(\sqrt{-11})$ isogénie près (ce qui suffit pour notre propos) équivaut alors à la détermination de la composante qui contient les invariants du normalisateur du groupe de Cartan non déployé.

Le chapitre I est consacré à l'étude des courbes elliptiques quotients de $J_o(121)$. Dans le §1, on détermine une base des formes paraboliques de poids 2 sur $\Gamma_o(121)$. Ceci n'est pas nécessaire pour la suite des calculs, mais il est commode d'avoir une expression permettant d'écrire des développements de ces formes. On donne ensuite dans le §2 la liste des courbes obtenues, ainsi que les propriétés de ces dernières. Après avoir rappelé au §3 les résultats sur lesquels se basent les calculs, on décrit ces derniers dans le §4. Les principaux résultats numériques sont rassemblés dans les tables. On explique au §5 comment on déduit du §4 les résultats annoncés, et on fait la liaison avec les tables de Swinnerton-Dyer et Vélu.

Le chapitre II développe la première des deux méthodes exposées plus haut. Les deux premiers §§ constituent des rappels sur les propriétés des courbes associées à un groupe de congruence de niveau p, pour p premier, $p \geqslant 5$. On s'intéresse en particulier au calcul du genre, et à la construction de fonctions au moyen de formes de Klein. Le §3 est également un rappel sur les façons de munir ces courbes d'une structure sur $\mathbb{Q}$. L'étude proprement dite des courbes $X_{\text{dép}}(11)$, $X_{\text{ndép}}(11)$, $X_{\mathfrak{S}_4}(11)$ fait l'objet du §4.

Enfin, le chapitre III reprend le problème du précédent en termes de représentations de Hecke. Les trois premiers §§ rappellent les résultats de Hecke. Le §4 en tire les conséquences qui nous intéressent. On obtient en particulier le résultat suivant, qui précise un énoncé de Hecke :

(3) La jacobienne $J(11)$ d'un modèle convenable sur $\mathbb{Q}$ de $X(11)$, courbe associée à $\Gamma(11)$, est isogène sur $\mathbb{Q}(\sqrt{-11})$ au produit de 26 courbes elliptiques.

Je tiens à exprimer ici mes remerciements pour leur aide à J.-P. Serre, dont les questions sont à l'origine de ce travail, ainsi qu'à B. Mazur, dont les suggestions m'ont été précieuses.

Je remercie également Mme Bonnardel, qui a bien voulu se charger de la frappe de mon manuscrit.

Notations

Ce travail est divisé en trois chapitres ; chaque chapitre est subdivisé en paragraphes. Les renvois internes à un même chapitre ne mentionnent pas le numéro de ce dernier.

Soit p un nombre premier. On note $\Gamma(p)$ le groupe des matrices de $SL_2(\mathbb{Z})$ dont la réduction modulo p est la matrice unité. Le quotient du demi-plan de Poincaré complété $\mathfrak{H}^*$ par l'action de $\Gamma(p)$ est une surface de Riemann compacte notée $X(p)$.

I. Courbes elliptiques quotients de $J_o(121)$

0. Notations.

0.1 Soit N un entier > 1 . On note $\Gamma_o(N)$ le sous-groupe de $SL_2(\mathbb{Z})$ formé des matrices $\begin{pmatrix} a & b \\ c & d \end{pmatrix}$ telles que $c \equiv 0 \pmod{N}$. On désigne par $X_o(N)$ la courbe projective et lisse, définie sur $\mathbb{Q}$, qui est associée à $\Gamma_o(N)$. Plus précisément, le corps des fonctions de $X_o(N)$ est $\mathbb{Q}(j, j_N)$, où $j(z)$ désigne la fonction invariant modulaire, et où $j_N(z) = j(Nz)$. La surface de Riemann $X_o(N)(\mathbb{C})$ s'identifie de façon canonique au quotient de $\mathfrak{H}^*$, demi-plan de Poincaré complété, par l'action naturelle de $\Gamma_o(N)$ (cf. [7], 1.2). On note $J_o(N)$ la jacobienne de $X_o(N)$.

0.2 Considérons la courbe $X_o(121)$. C'est une courbe algébrique de genre 6. On convient de plonger $X_o(121)$ dans sa jacobienne $J_o(121)$ en envoyant la pointe à l'infini de la courbe sur l'élément neutre de la variété abélienne (cf. [7], 1.2).

0.3 L'objet de ce chapitre est la détermination de toutes les courbes elliptiques définies sur $\mathbb{Q}$ qui sont quotients sur $\mathbb{Q}$ de $J_o(121)$.

1. Formes paraboliques de poids 2 sur $\Gamma_o(121)$.

1.1 On notera $\langle\Gamma_o(N), 2\rangle_o$ (N entier > 1) l'espace des formes paraboliques de poids 2 sur $\Gamma_o(N)$. Un élément f de cet espace se développe en série de Fourier :

$$f(z) = \sum_{n=1}^{\infty} a(n).q^n \quad , \quad q = e^{2\pi i z} .$$

Soit χ un caractère de Dirichlet. On note

$$f_\chi(z) = \sum_{n=1}^{\infty} \chi(n).a(n)q^n$$

la forme "tordue par χ ".

1.2 Notre première tâche consiste à déterminer une base de l'espace $<\Gamma_o(121),2>_o$.

Pour ce faire, soit $\eta(z)$ la forme modulaire de Dedekind (cf. [7], 3.1) et considérons

$$f_1(z) = \eta(z)^2\ \eta(11z)^2 = q \prod_{n=1}^{\infty} (1-q^n)^2\ (1-q^{11n})^2 .$$

Il est bien connu (cf. par exemple [7], prop. 3.1.1) que f_1 est la forme primitive normalisée (cf. [7], déf. 2.4.1) de $<\Gamma_o(11),2>_o$.

Par conséquent, $f_1(z)$ et $f_1(11z)$ engendrent dans $<\Gamma_o(121),2>_o$ le sous-espace des formes non primitives. Pour des raisons qui apparaitront plus tard (cf. II, 4.2.13), nous poserons :

$$f_2(z) = f_1(z) - 11f(11z) .$$

f_1 et f_2 sont normalisées et engendrent le sous-espace des formes non primitives.

Soit d'autre part :

$$g_1(z) = \eta(z)\eta(11z)^2\eta(121z) = q^6 \prod_{n=1}^{\infty} (1-q^n)(1-q^{11n})^2(1-q^{121n}) .$$

Il est facile de montrer (on peut utiliser par exemple la prop. 3.1.1 de [7]) que g_1 est un élément de $<\Gamma_o(121),2>_o$.

1.3 On note T_n $(n \geqslant 1)$ les opérateurs de Hecke, et $\mathcal{T}$ l'algèbre engendrée par les T_n .

Soit K le corps quadratique imaginaire $\mathbb{Q}(\sqrt{-11})$, et soit O_K l'anneau des entiers de K . On note χ_{11} le caractère de Legendre modulo 11. Si $\alpha \in O_K$, on s'autorisera à noter $\chi_{11}(\alpha)$ l'image par χ_{11} d'un représentant dans $\mathbb{Z}$ de la classe de α modulo $\sqrt{-11}$.

1.4 La réponse au problème posé en 1.2 est donnée par la proposition suivante :

<u>Proposition</u> 1.4.1 : <u>Les formes primitives normalisées de</u> $<\Gamma_o(121),2>_o$ <u>sont les formes</u> f_3, f_4, f_5, f_6 <u>définies comme suit</u> :

$$f_3 = -g_1 | (2+3T_2+2T_4+T_8) ,$$
$$f_4 = g_1 | (2-T_4-T_8) ,$$
$$f_5 = -g_1 | (2-T_4+T_8)$$
$$f_6(z) = \sum_{\substack{\alpha \in O_K \\ \chi_{11}(\alpha)=1}} \alpha . q^{N_{K/\mathbb{Q}}(\alpha)} .$$

<u>On a les identités</u> :

$$f_{1,\chi_{11}} = f_3 \ , \quad f_{3,\chi_{11}} = \frac{1}{11}(10f_1+f_2) \ ;$$
$$f_{4,\chi_{11}} = f_5 \ , \quad f_{5,\chi_{11}} = f_4 \ ;$$
$$f_{6,\chi_{11}} = f_6 \ .$$

1.4.2 Donnons les premiers termes du développement de Fourier de f_3, f_4, f_5, f_6 :

$$f_3(z) \equiv q+2q^2-q^3+2q^4+q^5-2q^6+2q^7-2q^9+2q^{10}-2q^{12} - 4q^{13}+4q^{14}-q^{15}-4q^{16}+2q^{17}-4q^{18} \pmod{q^{20}} ;$$

$$f_4(z) \equiv q+q^2+2q^3-q^4+q^5+2q^6-2q^7-3q^8+q^9+q^{10}-2q^{12} + q^{13}-2q^{14}+2q^{15}-q^{16}-5q^{17}+q^{18}+6q^{19} \pmod{q^{20}} ;$$

$$f_5(z) \equiv q-q^2+2q^3-q^4+q^5-2q^6+2q^7+3q^8+q^9-q^{10}-2q^{12} -q^{13}-2q^{14}+2q^{15}-q^{16}+5q^{17}-q^{18}-6q^{19} \pmod{q^{20}} ;$$

$$f_6(z) \equiv q-q^3-2q^4-3q^5-2q^9+2q^{12}+3q^{15}+4q^{16} \pmod{q^{20}} .$$

1.5 <u>Démonstration de la proposition</u> 1.4.1.

L'idée naturelle pour construire des formes paraboliques, connaissant l'une d'entre elles, est de faire opérer l'algèbre $\mathcal{T}$ sur cette dernière.

1.5.1 On part donc de g_1 et on essaie de déterminer $g_1|\mathcal{T}$. Pratiquement, on développe g_1 avec suffisamment de coefficients pour pouvoir à chaque étape appliquer le principe suivant, conséquence du théorème de Riemann-Roch :

Si les 11 premiers coefficients de f sont nuls, alors $f=0$.

On constate alors que f_1, f_2, g_1, g_2, g_3, où

$$g_2 = g_1|T_2 \quad , \quad g_3 = g_2|T_2 ,$$

engendrent dans $\langle\Gamma_o(121),2\rangle_o$ un sous-espace de codimension 1 qui est stable sous l'action de $\mathcal{T}$.

Il suffit alors de diagonaliser T_2, par exemple, dans ce sous-espace, pour obtenir f_3, f_4 et f_5. Par contre, l'invariance sous $\mathcal{T}$ montre qu'on ne peut pas espérer obtenir f_6 de cette façon.

Les formes f_3, f_4 et f_5 ont leurs coefficients entiers rationnels : Il en est donc de même des coefficients de la forme primitive normalisée manquante f_6.

1.5.2 La façon la plus simple de déterminer cette dernière consiste à vérifier que l'expression donnée par la proposition 1.4.1 définit bien une forme primitive normalisée, distincte de f_3, f_4 et f_5.

Pour ce faire, choisissons un plongement de K dans $\mathbb{C}$, et considérons l'unique Grössencharakter λ sur K, dont le conducteur est $\sqrt{-11}$, qui vérifie :

$$\lambda((\alpha)) = \alpha \qquad \text{pour} \quad \alpha \in 1 + \sqrt{-11}.O_K ,$$

et qui est à valeurs dans K. En d'autres termes :

$$\lambda((\alpha)) = \chi_{11}(\alpha).\alpha \quad \text{pour tout} \quad \alpha \in O_K .$$

Il résulte par exemple de [15] lemme 3, p. 203, que la forme $f_6(z)$ associée à λ est un élément de $\langle\Gamma_o(121),2\rangle_o$, vecteur propre de tous les T_n pour n premier à 11. Or cette forme n'est autre que celle considérée dans la proposition, et il est clair qu'elle est primitive et distincte de f_3, f_4 et f_5.

1.5.3 Considérons enfin l'opérateur R_{11}^* d'Atkin-Lehner ([1], 6.4). Nous verrons plus loin (3.4) comment cet opérateur s'interprète en termes géométriques. En tous cas, cet opérateur transforme un sous-espace propre sous $\mathcal{T}$ en un autre sous-espace propre.L'examen des premiers coefficients des formes f_i, $i = 1,\dots,6$, montre que l'on a :

$$f_1|R_{11}^* = i\sqrt{11}.f_{1,\chi_{11}} = i\sqrt{11}\ f_3 ,$$

$$f_4|R_{11}^* = i\sqrt{11}.f_{4,\chi_{11}} = i\sqrt{11}.f_5 ,$$

$$f_6|R_{11}^* = i\sqrt{11}.f_{6,\chi_{11}} = i\sqrt{11}.f_6 .$$

2. Enoncé des résultats.

L'objet de ce paragraphe est d'énoncer les résultats obtenus concernant les courbes elliptiques quotients de $J_o(121)$. Ces résultats sont rassemblés dans la proposition 2.6.1. La démonstration de cette dernière occupe les trois paragraphes suivants.

2.1 Considérons les six formes normalisées déterminées au paragraphe 1 :

$$f_i(z) = \sum_{n=1}^{\infty} a_i(n).q^n \quad , i = 1,\ldots,6 .$$

Les coefficients de chacune de ces formes sont des entiers rationnels. De plus, ces formes sont des vecteurs propres de tous les opérateurs de Hecke T_n $(n \geqslant 1)$. D'après Shimura ([13], th. 1), il existe des morphismes canoniques :

$$\varphi_i : J_o(121) \to E_i \quad , i = 1,\ldots,6 ,$$

définis sur $\mathbb{Q}$, dont l'image est une courbe elliptique E_i définie sur $\mathbb{Q}$. Moyennant l'application canonique du demi-plan de Poincaré $\mathfrak{H}^*$ dans $X_o(121)$, l'image inverse par φ_i d'une différentielle de première espèce non nulle sur E_i est un multiple non nul de $f_i(z)dz$. Enfin, le noyau de φ_i est une variété abélienne. En particulier, φ_i est une paramétrisation de Weil de E_i , et E_i est une courbe de Weil (cf. [10]) pour $i = 3,4,5,6$.

On aura besoin plus loin (cf. §3) d'une description plus précise des morphismes φ_i .

2.2 Soit C_J le groupe engendré dans la jacobienne $J_o(121)$ par les pointes de $X_o(121)$. C'est un groupe rationnel sur $\mathbb{Q}$ (le groupe de Galois agit sur les pointes par permutation, cf. II, §3), d'ordre fini.

(Plus précisément, on sait que l'ordre d'une quelconque des pointes divise 25.11, d'après [8]).

On désigne par C_i $(i = 1,\ldots,6)$ l'image du groupe C_J dans la courbe E_i. Nous dirons que le groupe C_J, resp. C_i, est le groupe cuspidal de $J_o(121)$, resp. E_i.

2.3 Soit E une courbe elliptique définie sur $\mathbb{Q}$. On suppose que E n'a pas d'autres automorphismes que ± 1. Alors, il existe, à isomorphisme près sur $\mathbb{Q}$, une courbe E' et une seule, telle que E' soit isomorphe à E sur $\mathbb{Q}(\sqrt{-11})$ mais non sur $\mathbb{Q}$. On dit que E' est la forme tordue de E sur $\mathbb{Q}(\sqrt{-11})$.

2.4 Soit $n \mapsto [n]$ l'isomorphisme canonique de $\mathbb{F}_{11}^{\times}$ avec le groupe de Galois de $\mathbb{Q}(\zeta_{11})$ sur $\mathbb{Q}$, ζ_{11} désignant une racine primitive 11-ième de l'unité.

On note

(2.4.1)
$$\varepsilon : \mathbb{F}_{11}^{\times} \to (\mathbb{Z}/25\mathbb{Z})^{\times}$$

le morphisme de groupe multiplicatif qui est injectif et envoie la classe de 5 sur celle de -4. Ce morphisme induit un isomorphisme de $\mathbb{F}_{11}^{\times}$ avec le sous-groupe de $(\mathbb{Z}/25\mathbb{Z})^{\times}$ formé des éléments qui sont des carrés.

2.5 On se permettra de noter $\mu_{11}^{\otimes a}$ (a un entier mod 10) le $\mathrm{Gal}(\mathbb{Q}(\zeta_{11})/\mathbb{Q})$-module galoisien défini de la façon suivante :

le groupe sous-jacent à $\mu_{11}^{\otimes a}$ est le groupe cyclique $\mathbb{Z}/11\mathbb{Z}$, noté additivement ; l'élément $[n]$ de $\mathrm{Gal}(\mathbb{Q}(\zeta_{11})/\mathbb{Q})$ agit sur $P \in \mathbb{Z}/11\mathbb{Z}$ par :

$$P^{[n]} = n^a.P .$$

2.6 Nous pouvons maintenant énoncer le résultat principal de ce paragraphe :

<u>Proposition</u> 2.6.1 : 1) <u>Il existe douze courbes elliptiques définies sur</u> $\mathbb{Q}$ <u>qui sont quotients sur</u> $\mathbb{Q}$ <u>de</u> $J_o(121)$. <u>Trois d'entre elles ont pour conducteur</u> 11 : <u>ce sont les courbes</u> E_1, E_2, E_2/C_2. <u>Elles sont</u>

<u>liées par des isogénies rationnelles sur</u> $\mathbb{Q}$, <u>de degré</u> 5 :

(a) $E_2 \to E_1 \to E_2/C_2$;

<u>on a</u> $E_1 = E_2/5C_2$.

<u>Les neuf courbes restantes ont pour conducteur</u> 121. <u>Ce sont les quatre courbes</u> E_3 , E_4 , E_5 , E_6 , <u>et les courbes obtenues par les</u> $\mathbb{Q}$-<u>isogénies suivantes</u> :

(a') $E_3 \to E_3/5C_3 \to E_3/C_3$, <u>isogénies de degré</u> 5 ;

(b) $E_4 \to E_4/C_4$,

(c) $E_5 \to E_5/C_5$,

(d) $E_6 \to E_6/C_6$, <u>isogénies de degré</u> 11.

2) <u>On a</u> :

$$E_2' \cong E_3 \ , \ E_1' \cong E_3/5C_3 \ , \ (E_2/C_2)' \cong E_3/C_3 \ ,$$
$$E_4' \cong E_5/C_5 \ ,$$
$$E_5' \cong E_4/C_4 \ ,$$
$$E_6' \cong E_6/C_6 \ .$$

3) <u>La structure galoisienne des groupes cuspidaux</u> C_i $(i = 1, \dots, 6)$ <u>est la suivante</u> :

(a) C_2 <u>est cyclique d'ordre</u> 25, <u>et l'action du groupe de Galois est donnée par</u> :

<u>si</u> $P \in C_2$, $[n] \in \mathrm{Gal}(\mathbb{Q}(\zeta_{11})/\mathbb{Q})$,

<u>alors</u> $P^{[n]} = \chi_{11}(n).\varepsilon(n)P$;

$C_1 = C_2/5C_2$ <u>est un groupe cyclique d'ordre</u> 5 <u>muni de l'action triviale du groupe de Galois</u> ;

(a') C_3 <u>est cyclique d'ordre</u> 25, <u>et l'action du groupe de Galois est donnée par</u> :

<u>si</u> $P \in C_3$, $[n] \in \mathrm{Gal}(\mathbb{Q}(\zeta_{11})/\mathbb{Q})$,

<u>alors</u> $P^{[n]} = \varepsilon(n).P$;

(b) C_4 <u>est isomorphe à</u> $\mu_{11}^{\otimes 7}$ <u>en tant que module galoisien</u> ;

(c) C_5 <u>est isomorphe à</u> $\mu_{11}^{\otimes 9}$;

(d) C_6 <u>est isomorphe à</u> $\mu_{11}^{\otimes 8}$.

4) Les douze courbes considérées coïncident avec les courbes figurant dans la liste de Swinnerton-Dyer et Vélu (cf. [2], p. 82 et p. 97). Plus précisément, le tableau suivant donne la correspondance entre les courbes considérées ici et celle de la table 1 de [2] :

	conducteur 11			conducteur 121								
courbe	E_2	E_1	E_2/C_2	E_3	$E_3/5C_3$	E_3/C_3	E_6	E_6/C_6	E_4	E_4/C_4	E_5	E_5/C_5
notation dans [2] table 1	A	B	C	A	B	C	D	E	F	G	H	I

En particulier (cf. [16]) :

$$j(E_5) = -11.131^3 ,$$
$$j(E_4) = -11^2 ,$$
$$j(E_6) = -2^{15} .$$

Corollaire. Les coefficients $a_i(n)$ des formes f_i $(i=1,3,4,5,6)$ vérifient les congruences suivantes :

(a) $a_1(\ell) \equiv \chi_{11}(\ell)(\varepsilon(\ell)+\varepsilon(\ell)^{-1})$ (mod 25) ;

(a') $a_3(\ell) = \chi_{11}(\ell)a_1(\ell) \equiv \varepsilon(\ell)+\varepsilon(\ell)^{-1}$ (mod 25) ;

(b) $a_4(\ell) \equiv \ell^4+\ell^7$ (mod 11) ;

(c) $a_5(\ell) \equiv \ell^2+\ell^9$ (mod 11) ;

(d) $a_6(\ell) \equiv \ell^3+\ell^8$ (mod 11) .

pour tout ℓ premier $\neq 11$.

La démonstration de la proposition 2.6.1 va faire l'objet des paragraphes 3, 4 et 5.

3. Rappel de résultats généraux.

Ce paragraphe regroupe un certain nombre de résultats de nature générale dont nous aurons besoin dans les paragraphes 4 et 5.

Il s'agit tout d'abord de la description, en termes de périodes, des variétés abéliennes associées aux formes paraboliques normalisées vecteurs propres des opérateurs de Hecke. Cette description est due à

Shimura (cf. [13], [14] et [15]). On la rappelle en 3.1.

On utilise également les propriétés des symboles modulaires de Birch et Manin (cf. [9] et [10]). En particulier, ces symboles permettent de décrire l'homologie de $X_o(N)$; cf. 3.2. L'action des opérateurs de Hecke se traduit par des identités rappelées en 3.3.

Enfin, on fait un rappel rapide des définitions et de quelques propriétés de certains endomorphismes de torsion analogues à ceux utilisés par Shimura dans [15]. Ceci fait l'objet de 3.4.

Dans tout ce paragraphe, on note N un entier >1, et on suppose choisie une base $f_1,\dots,f_{g_J}$ de l'espace $\langle\Gamma_o(N),2\rangle_o$ (on note g_J le genre de $X_o(N)$). Soit $J_o(N) = J$ la jacobienne de $X_o(N)$.

3.1 <u>Périodes des formes paraboliques</u>.

L'espace $\langle\Gamma_o(N),2\rangle_o$ s'identifie de façon canonique avec l'espace D_J des 1-formes différentielles invariantes sur J. Soit Ω_f la forme différentielle associée à f. Le choix de la base $\{f_1,\dots,f_{g_J}\}$ permet d'identifier à $\mathbb{C}^{g_J}$ l'espace tangent à J à l'origine $\mathrm{Hom}_{\mathbb{C}\text{-lin}}(D_J,\mathbb{C})$:

(3.1.1) $$\mathrm{Hom}_{\mathbb{C}\text{-lin}}(D_J,\mathbb{C}) \xrightarrow{\sim} \mathbb{C}^{g_J},$$

en associant à une forme linéaire ses valeurs sur $\Omega_{f_1},\dots,\Omega_{f_{g_J}}$.

Soient z_o, z_1 deux éléments du demi-plan de Poincaré complété $\mathfrak{H}^* = \mathfrak{H} \cup \mathbb{Q} \cup \{i\infty\}$. On note

$$\{z_o,z_1\} \in \mathbb{C}^{g_J},$$

le vecteur de coordonnées

$$\{z_o,z_1\}_{f_i} = \int_{z_o}^{z_1} f_i(z)dz \qquad i = 1,\dots,g_J .$$

On peut alors énoncer ([14], §3) :

<u>Proposition</u> 3.1.2 : <u>Soit</u> L_J <u>le sous-</u>$\mathbb{Z}$<u>-module de</u> $\mathbb{C}^{g_J}$ <u>engendré par les vecteurs</u> $\{z_o,M(z_o)\}$, <u>où</u> $z_o \in \mathfrak{H}^*$ <u>est fixé, et où</u> M <u>parcourt</u> $\Gamma_o(N)$.

<u>L'application</u>

$$\Phi : \mathfrak{h}^* \to \mathbb{C}^{g_J}$$
$$z \mapsto \left(\int_{i\infty}^{z} f_i(t)dt\right)_{i=1,\dots,g_J}$$

<u>induit un isomorphisme de variétés abéliennes</u> :

$$\Phi : J(\mathbb{C}) \xrightarrow{\sim} \mathbb{C}^{g_J}/L_J .$$

Considérons maintenant une forme $f \in \langle\Gamma_o(N),2\rangle_o$ qui soit vecteur propre de tous les T_n pour n premier à N, et normalisée ([7], déf. 2.1.1). Soit $\mathcal{F}$ le sous-corps de $\mathbb{C}$ engendré sur $\mathbb{Q}$ par les coefficients de f. Enfin, soit I l'ensemble des plongements de $\mathcal{F}$ dans $\mathbb{C}$.

Alors, à la forme f est associé de façon canonique un morphisme $\mathbb{Q}$-rationnel surjectif

$$\varphi_A : J \to A$$

où A est une variété abélienne sur $\mathbb{Q}$, de dimension $g_A = [\mathcal{F}:\mathbb{Q}]$. De plus ([14], prop. 3) :

<u>Proposition</u> 3.1.3 : <u>Soit</u> $f_1,\dots,f_{g_A}$ <u>une base de</u> $\bigoplus_{\sigma\in I} \mathbb{C}.f_\sigma$, <u>où</u> $f_\sigma(z) = \sum_{n=1}^{\infty} a(n)^\sigma.q^n$. <u>Soit</u> L_A <u>le sous-</u>$\mathbb{Z}$<u>-module de</u> $\mathbb{C}^{g_A}$ <u>engendré par les vecteurs</u>

$$\{z_o,M(z_o)\}_{f_i}, \qquad i = 1,\dots,g_A ,$$

<u>pour</u> $z_o \in \mathfrak{h}^*$ <u>fixé, et</u> M <u>parcourant</u> $\Gamma_o(N)$.

<u>Alors</u> Φ <u>induit un isomorphisme</u> Φ_A <u>qui rend le diagramme suivant commutatif</u> :

$$\begin{array}{ccc} J(\mathbb{C}) & \xrightarrow{\Phi} & \mathbb{C}^{g_J}/L_J \\ {\scriptstyle\tilde\varphi_A}\downarrow & & \downarrow \\ A(\mathbb{C}) & \xrightarrow{\Phi_A} & \mathbb{C}^{g_A}/L_A \end{array}$$

<u>la flèche verticale de droite étant celle qui provient de la projection</u> $\mathbb{C}^{g_J} \to \mathbb{C}^{g_A}$ <u>sur les</u> g_A <u>premières composantes</u>.

3.2 <u>Description de l'homologie en termes de symboles modulaires</u>.

Considérons le groupe d'homologie $H_1(X_o(N)(\mathbb{C}),\mathbb{Z})$ de la surface de Riemann $X_o(N)(\mathbb{C})$. Il existe une façon naturelle de l'identifier à un réseau de rang maximal dans l'espace tangent à J à l'origine grâce à la flèche :

$$H_1(X_o(N)(\mathbb{C}),\mathbb{Z}) \to \mathrm{Hom}_{\mathbb{C}\text{-lin}}(D_J,\mathbb{C})$$

définie par

$$\beta \mapsto (\Omega \mapsto \int_\beta \Omega) .$$

Par composition avec l'isomorphisme (3.1.1), on obtient une injection de $H_1(X_o(N)(\mathbb{C}),\mathbb{Z})$ dans $\mathbb{C}^{g_J}$, définie par

(3.2.1) $$\beta \mapsto (\int_\beta f_i(z)dz)_{i=1,\ldots,g_J} \in \mathbb{C}^{g_J} .$$

Par tensorisation avec $\mathbb{R}$ cette injection donne un isomorphisme d'espaces vectoriels réels.

Rappelons maintenant le résultat suivant de Manin ([9], prop. 1.4):

<u>Proposition</u> 3.2.2 : <u>Soit</u> $z_o \in \mathfrak{h}^*$. <u>L'application</u>

$$\Gamma_o(N) \to H_1(X_o(N)(\mathbb{C}),\mathbb{Z})$$
$$M \mapsto \left\{\begin{matrix}\text{classe d'un chemin}\\ \text{joignant } z_o \text{ à } M(z_o)\end{matrix}\right\}$$

<u>est un homomorphisme surjectif de groupes</u>, <u>indépendant du choix de</u> z_o .

Par composition avec (3.2.1), on obtient l'application $\Gamma_o(N) \to \mathbb{C}^{g_J}$ définie par

$$M \mapsto \{z_o,M(z_o)\} .$$

Vu la définition du réseau L_J (cf. prop. 3.1.2), ceci montre que le réseau L_J s'identifie à l'image de $H_1(X_o(N)(\mathbb{C}),\mathbb{Z})$ par l'injection (3.2.1).

Il reste à décrire le réseau L_J en termes de symboles modulaires. Ceci est réalisé par la proposition suivante ([9] th. 2.7, a) :

<u>Proposition</u> 3.2.3 : <u>Soit</u> $\aleph'(N)$ <u>le groupe abélien engendré par les symboles</u> $(\tilde{c}:\tilde{d}) \in \mathbb{P}_1(\mathbb{Z}/N\mathbb{Z})$, <u>avec les relations</u> :

$$(\tilde{c}:\tilde{d}) + (-\tilde{d}:\tilde{c}) = 0$$
$$(\tilde{c}:\tilde{d}) + ((\tilde{c}-\tilde{d}):\tilde{c}) + (-\tilde{d}:(\tilde{c}-\tilde{d})) = 0 .$$

<u>Soit</u> $\aleph(N)$ <u>le sous-groupe de</u> $\aleph'(N)$ <u>noyau de l'opération cobord</u> (cf. [9], 2.6).

<u>Soit enfin</u>

$$\xi((\tilde{c}:\tilde{d})) = \{\frac{b}{d}, \frac{a}{c}\} \text{ , où } a,b,c,d \in \mathbb{Z}$$

<u>vérifient</u> $ad-bc = 1$, <u>et</u> $\tilde{c} \equiv c(\mathrm{mod}\ N)$, $\tilde{d} \equiv d(\mathrm{mod}\ N)$.

<u>Alors l'application</u> ξ <u>se prolonge par linéarité en un isomorphisme de groupes abéliens</u> :

$$\xi : \aleph(N) \to L_J .$$

La proposition suivante ([9], th. 2.7, b) permet de ramener la détermination de $\{0, \frac{b}{a}\}$, $\frac{b}{a} \in \mathbb{Q}$, à celle d'une somme d'éléments contenus dans l'image de ξ :

<u>Proposition</u> 3.2.4 : <u>Soit</u> $\frac{b}{a} \in \mathbb{Q}$, $\frac{b}{a} > 0$. <u>Considérons les convergents successifs du développement de</u> $\frac{b}{a}$ <u>en fraction continue</u> :

$$\frac{b}{a} = \frac{b_n}{a_n} , \frac{b_{n-1}}{a_{n-1}}, \dots, \frac{b_1}{a_1} , \frac{b_0}{a_0} = \frac{b_0}{1} .$$

<u>Alors</u>, <u>on a</u>

$$\{0, \frac{b}{a}\} = \sum_{k=1}^{n} \xi((-1)^{k-1}\tilde{a}_k : \tilde{a}_{k-1}) .$$

Revenons à l'identification des points de $J(\mathbb{C})$ avec le quotient de $\mathbb{C}^{g_J}$ par L_J :

$$\Phi : J(\mathbb{C}) \to \mathbb{C}^{g_J}/L_J .$$

La courbe $X_0(N)$ est définie sur $\mathbb{Q}$, de façon canonique. Par conséquent, la conjugaison complexe opère sur $X_0(N)(\mathbb{C})$. Cette action n'est autre que la symétrie par rapport à l'axe imaginaire, lorsqu'on identifie $X_0(N)(\mathbb{C})$ à $\mathfrak{H}^*/\Gamma_0(N)$ de la façon canonique ([9], § 2.1).

Il en résulte une action sur l'homologie, donc sur L_J . En particulier, l'action sur les éléments contenus dans l'image de ξ est la suivante ([9], cor. 2.5) :

<u>Proposition</u> 3.2.5 : <u>La conjugaison complexe opère de la façon suivante sur</u> L_J :

$$\overline{\xi(\tilde{c}:\tilde{d})} = -\xi(\tilde{d}:\tilde{c}) .$$

3.3 <u>Opérateurs de Hecke et symboles modulaires</u>.

Supposons maintenant que la base $f_1,\dots,f_{g_J}$ choisie au début de ce paragraphe soit formée de vecteurs propres pour les opérateurs de Hecke T_n , pour $(n,N) = 1$. On a donc

$$f_i \mid T_n = c_i(n).f_i \qquad \begin{array}{l} i = 1,\dots,g_J \\ n \geqslant 1 \\ (n,N) = 1 . \end{array}$$

On note $\underline{c}(n)$ la matrice diagonale

$$\underline{c}(n) = \begin{pmatrix} c_1(n) & & 0 \\ & \ddots & \\ 0 & & c_{g_J}(n) \end{pmatrix}$$

Le théorème 3.5 de [9] s'exprime alors de la façon suivante :

<u>Proposition</u> 3.3.1 : <u>Soit</u> $z \in \mathfrak{h}^*$. <u>Sous les hypothèses précédentes</u>, <u>on a</u> :

(3.3.1.1) $$\underline{c}(n).\{0,z\} = \sum_{\substack{d\mid n \\ b \bmod d}} \left(\{0,\frac{n d^{-1}z+b}{d}\}-\{0,\frac{b}{d}\}\right) .$$

(On identifie les éléments de $\mathbb{C}^{g_J}$ à des vecteurs à g_J colonnes).

C'est cette formule qui nous permettra de déterminer la structure du groupe engendré par les pointes.

<u>Remarque</u> 3.3.2 : 1) Choisissons $2g_J$ valeurs de $z \in \mathfrak{h}^*$ de telle sorte que les vecteurs $\{0,z\}$ engendrent le réseau L_J . Il est clair que les vecteurs figurant au second membre de (3.3.1.1) sont des éléments de L_J . On obtient donc de cette façon une représentation entière de dimension $2g_J$.

2) Soit ℓ un nombre premier, premier à N. Soit $z = \frac{u}{v}$ une pointe de $X_o(N)$, v étant un diviseur positif de N, et u un entier défini modulo le pgcd de v et N/v (cf. [9], prop. 2.2). La formule (3.3.1.1), appliquée à

$$z = \frac{u}{v} \quad \text{et} \quad n = \ell$$

devient :

$$(3.3.2.1) \qquad (\ell+1-\underline{c}(\ell))\{0,\tfrac{u}{v}\} = \{0,\tfrac{u}{v}\} - \{0,\tfrac{\ell u}{v}\} + \sum_{b=0}^{\ell-1} (\{0,\tfrac{u}{v}\} - \{0,\tfrac{u+bv}{\ell v}\}) + \sum_{b=0}^{\ell-1} \{0,\tfrac{b}{\ell}\} .$$

Si on suppose de plus que $\ell \equiv 1 \pmod{(v,N/v)}$, le second membre est un élément de L_J. On en déduit que l'ordre de la différence des pointes 0 et $\frac{u}{v}$ dans la jacobienne divise $\det(1+\ell-\underline{c}(\ell))$, pour tout $\ell \equiv 1 \pmod{(v,N/v)}$.

3.4 <u>Endomorphismes de torsion</u>. On suppose, dans cette partie seulement, que $N = p^2$, où p est premier > 3.

Considérons la courbe $X_o(p^2)$, pour p premier > 3.

La matrice $\begin{pmatrix} 1 & u/p \\ 0 & 1 \end{pmatrix}$, pour u entier premier à p, définit une correspondance modulaire sur la courbe $X_o(p^2)$, ainsi qu'il est expliqué dans [13], 7.3. Cette correspondance ne dépend que du résidu quadratique de $u \pmod p$. On la notera $R_p^{\pm}$ selon que $(\frac{u}{p}) = \pm 1$. On note de même les endomorphismes de la jacobienne $J_o(p^2)$ définis par $R_p^{\pm}$.

Soit $R_p = R_p^+ - R_p^-$.

On vérifie sans peine les propriétés suivantes (cf. [15]) :

(i) R_p est définie sur le corps $\mathbb{Q}(\sqrt{(\frac{-1}{p})p})$;

(ii) $R_p^3 = (\frac{-1}{p}).p\, R_p$;

(iii) $R_p \circ T_\ell = (\frac{\ell}{p})\, T_\ell \circ R_p$, pour tout ℓ premier distinct de p.

L'espace $\langle\Gamma_o(p^2),2\rangle_o$ s'identifie de façon canonique avec l'espace des 1-formes différentielles invariantes sur $J_o(p^2)$. L'endomorphisme

R_p induit donc un opérateur sur $<\Gamma_o(p^2),2>_o$, et l'on a :

(iv) L'opérateur induit par R_p sur $<\Gamma_o(p^2),2>_o$ est l'opérateur R_p^* d'Atkin et Lehner [1], 6.4, défini par :

$$(f|R_p^*)(z) = \sum_{a \bmod p} (\tfrac{a}{p}) f(z+\tfrac{a}{p})$$
$$= \begin{cases} \sqrt{p}.f_\chi(z) & \text{si} \quad p \equiv 1 \pmod 4, \\ i\sqrt{p}.f_\chi(z) & \text{si} \quad p \equiv 3 \pmod 4. \end{cases}$$

(χ désigne le caractère de Legendre modulo p).

4. Description des calculs.

4.1 Dans ce paragraphe, nous allons utiliser les résultats généraux rappelés dans le paragraphe précédent.

Reprenons donc les notations du §3, et supposons que $N = 121$. Soit $f_1,\dots,f_6$ la base de $<\Gamma_o(121),2>_o$ déterminée au §1.

4.2 Dans un premier temps (cf. 4.6), nous décrivons le réseau $L = L_J$ en termes des symboles modulaires introduits en 3.2. Nous en ferons de même pour les réseaux L^+ , L^- correspondant aux valeurs propres $+1$ et -1 de la conjugaison complexe (cf. (4.7)). Ceci fait, la prop. 3.1.3 nous permettra de décrire en termes de réseaux les morphismes $\varphi_i : J_o(121) \to E_i$ associés aux formes f_i $(i = 1,\dots,6)$ (cf. 4.8).

4.3 Dans un second temps (cf. (4.9)), nous utilisons la description de l'action des opérateurs de Hecke donnée par la prop. 3.3.1. Plus précisément, faisons parcourir à $z \in \mathfrak{H}^*$ un ensemble de valeurs tel que $\{0,z\} \pm \{0,-z\}$ engendre le réseau L^+, resp. L^- (cf. rem. 3.3.2), et utilisons la prop. 3.3.1, pour un nombre premier $n = \ell$ fixé (en fait, $n = 2$). Le système linéaire obtenu permet d'une part, de déterminer les valeurs propres de T_n , et d'autre part, de déterminer les coordonnées des générateurs de L^+ resp. L^- en fonction de générateurs ω_i^+ , resp. ω_i^- des projections L_i^+ , L_i^- , $i = 1,\dots,6$ des réseaux L^+ resp. L^- .

4.4 On peut alors utiliser les résultats obtenus pour déterminer l'ordre des pointes de $X_o(121)$. C'est ce qui est fait dans 4.10. Pour cela, on utilise de nouveau la proposition 3.3.1.

4.5 Enfin, on montre dans 4.11 quelles sont les relations correspondant aux identités $f_{1,\chi_{11}} = f_3$, etc. du §1.

4.6 <u>Le groupe</u> $\mathfrak{H}(121)$.

Considérons tout d'abord l'ensemble $\mathbb{P}^1(\mathbb{Z}/121\mathbb{Z})$. Les éléments de cet ensemble sont les couples suivants :

$$(\tilde{1}:\tilde{0})\ ,\ (\tilde{0};\tilde{1})\ ;\ (\tilde{c},\tilde{1})\ ,\ c=0,\dots,120\ ;\ (\tilde{c}:\widetilde{11})\ ,\ c=1,\dots,10\ .$$

Avec les notations de Manin [9], prop. 2.2 , les douze pointes de $X_o(121)$ sont :

$$[1:1 \bmod 1]\ ,\ [121;1 \bmod 1]\ ,\ [11;d \bmod 11]\ ,\ d=1,\dots,10\ .$$

L'application cobord ([9], cor. 2.6) est la suivante :

$$\partial(\tilde{1}:\tilde{0}) = [1:1 \bmod 1]-[121;1 \bmod 1]\ ;$$

$$\partial(\tilde{0}:\tilde{1}) = [121;1 \bmod 1]-[1;1 \bmod 1]\ ;$$

$$\partial(\tilde{c}:\tilde{1}) = \begin{cases}[11;\frac{c}{11} \bmod 11]-[1;1 \bmod 1] & \text{si } 11 \text{ divise } c\ ,\\ 0 \quad \text{sinon} & (c=1,\dots,120);\end{cases}$$

$$\partial(\tilde{c}:\widetilde{11}) = [1;1 \bmod 1]-[11;-c^{-1} \bmod 11],\ (c=1,\dots,10).$$

Les 110 éléments $(\tilde{c}:\tilde{1})$, $c = 1,\dots,120$, $(c,11) = 1$, sont donc des éléments de $\mathfrak{H}(121)$. De plus, leurs images engendrent $\mathfrak{H}(121)$, comme le montre la prop. 3.2.3. La même proposition permet, après un calcul facile mais fastidieux, d'obtenir un système de générateurs :

<u>Lemme</u> 4.6.1 : <u>Le groupe</u> $\mathfrak{H}(121)$ <u>admet comme système de générateurs les douze éléments suivants</u> :

$$(\tilde{c},\tilde{1})\ ,\ c \in \mathcal{B}\ ,$$

<u>où</u> $\mathcal{B} = \{2,3,4,5,7,8,9,18,26,36,49,51\}$.

Tout élément de $\mathfrak{H}(121)$ s'écrit donc comme combinaison linéaire à coefficients entiers de ces douze éléments. La table 1 donne, à toutes fins utiles, cette expression pour chacun des éléments $(\widetilde{c}:\widetilde{1})$, $c = 1,\dots,120$, $(c,11) = 1$.

D'après la prop. 3.2.3, le réseau L_J est l'image du groupe $\aleph(121)$ par l'application ξ . Tenant compte de ce que

$$\xi(\widetilde{c}:\widetilde{1}) = \{0,\tfrac{1}{c}\} \text{ , pour tout } c \in \mathbb{Z} \text{ ,}$$

le lemme 4.6.1 s'énonce sous la forme équivalente :

<u>Proposition</u> 4.6.2 : <u>Le réseau</u> $L = L_J$ <u>est le réseau engendré dans</u> $\mathbb{C}^6$ <u>par les douze vecteurs</u> :

$$\{0,\tfrac{1}{2}\} , \{0, \tfrac{1}{3}\} , \{0, \tfrac{1}{4}\} , \{0, \tfrac{1}{5}\} , \{0, \tfrac{1}{7}\} , \{0, \tfrac{1}{8}\} ,$$
$$\{0,\tfrac{1}{9}\} , \{0,\tfrac{1}{18}\} , \{0,\tfrac{1}{26}\} , \{0,\tfrac{1}{36}\} , \{0,\tfrac{1}{49}\} , \{0,\tfrac{1}{51}\} .$$

4.7 <u>Les réseaux</u> L^+ <u>et</u> L^- .

Introduisons maintenant l'action de la conjugaison complexe. Le conjugué de chacun des vecteurs qui engendrent L est donné par la prop. 3.2.5. L'utilisation de la prop. 3.2.3 et de la table 1 permet d'exprimer ce transformé en fonction des générateurs. Les résultats obtenus sont donnés dans la table 2.

Un calcul facile permet de déterminer les sous-réseaux L^+ et L^- associés aux valeurs propres $+1$ et -1. On trouve :

<u>Proposition</u> 4.7.1 : <u>Soient</u> $L^{\pm}$ <u>les sous-réseaux de</u> L <u>correspondant aux valeurs</u> ± 1 <u>de la conjugaison complexe</u> (prop. 3.2.5). <u>Alors</u> :

L^+ <u>est engendré par les vecteurs</u> :

$$\gamma_1^+ = \{0, \tfrac{1}{3}\} + \{0,-\tfrac{1}{3}\} , \quad \gamma_2^+ = \{0,\tfrac{1}{36}\} + \{0,-\tfrac{1}{36}\} ,$$
$$\gamma_3^+ = \{0, \tfrac{1}{8}\} + \{0,-\tfrac{1}{8}\} , \quad \gamma_4^+ = \{0,\tfrac{1}{51}\} + \{0,-\tfrac{1}{51}\} ,$$
$$\gamma_5^+ = \{0,\tfrac{1}{26}\} + \{0,-\tfrac{1}{26}\} , \quad \gamma_6^+ = \{0,\tfrac{1}{18}\} + \{0,-\tfrac{1}{18}\} ,$$

L^- <u>est engendré par les vecteurs</u> :

$$\gamma_1^- = \{0, \tfrac{1}{3}\} - \{0,-\tfrac{1}{3}\} , \quad \gamma_2^- = \{0, \tfrac{1}{9}\} - \{0,-\tfrac{1}{9}\} ,$$
$$\gamma_3^- = \{0, \tfrac{1}{5}\} - \{0,-\tfrac{1}{5}\} , \quad \gamma_4^- = \{0, \tfrac{1}{7}\} - \{0,-\tfrac{1}{7}\} ,$$
$$\gamma_5^- = \{0,\tfrac{1}{49}\} - \{0,-\tfrac{1}{49}\} , \quad \gamma_6^- = \{0,\tfrac{1}{36}\} - \{0,-\tfrac{1}{36}\} .$$

La table 3 donne l'expression des vecteurs $\{0,\frac{1}{c}\}$, $c \in \mathfrak{B}$, qui engendrent L, en fonction des vecteurs $\gamma_k^{\pm}$, $k=1,\ldots,6$. Un calcul élémentaire permet de montrer le résultat suivant :

Proposition 4.7.2 : Le réseau L est un sous-réseau d'indice 2^6 du réseau $\frac{1}{2}L^+ \oplus \frac{1}{2}L^-$. Plus précisément, un élément z de ce dernier .

$$z = \sum_{k=1}^{6} \left(\frac{a_k}{2}\gamma_k^+ + \frac{b_k}{2}\gamma_k^-\right) , \qquad a_k, b_k \in \mathbb{Z} ,$$

est un élément de L si et seulement si les conditions suivantes sont vérifiées :

$$\begin{aligned} b_1 &\equiv a_1 + a_3 + a_5 \pmod 2, \\ b_2 &\equiv a_5 \pmod 2, \\ b_3 &\equiv a_3 + a_4 + a_5 + a_6 \pmod 2, \\ b_4 &\equiv a_4 + a_6 \pmod 2, \\ b_5 &\equiv a_6 \pmod 2, \\ b_6 &\equiv a_2 \pmod 2). \end{aligned}$$

4.8 Les courbes E_i $(i = 1,\ldots,6)$.

Soit f_i l'une des formes paraboliques déterminée au §1. La proposition 3.1.3 s'applique à f_i et donne :

Proposition 4.8.1 : Soit $\pi_i : \mathbb{C}^6 \to \mathbb{C}$ la i-ième projection $(i = 1,\ldots,6)$. Soit L_i le sous-réseau de $\mathbb{C}$ image de L par π_i :

$$L_i = \pi_i(L) .$$

L'application :

$$\Phi : \mathfrak{H}^* \to \mathbb{C}^6$$
$$z \mapsto \{i\infty, z\} = \left(\int_{i\infty}^{z} f_i(t)dt\right)_{i=1,\ldots,6}$$

induit des isomorphismes Φ , Φ_i de variétés abéliennes rendant commutatifs les diagrammes suivants :

$$\begin{array}{ccc} J_0(121)(\mathbb{C}) & \xrightarrow{\Phi} & \mathbb{C}^6/L \\ \varphi_i \downarrow & & \downarrow \pi_i \\ E_i(\mathbb{C}) & \xrightarrow{\Phi_i} & \mathbb{C}/L_i \end{array} .$$

4.8.2 Notations. On pose : $L_i^{\pm} = \pi_i(L^{\pm})$.

Soit $z \in \mathfrak{h}^*$. On convient de poser :

$$\{0,z\} \pm \{0,-z\} = \sum_{k=1}^{6} x^{\pm}(z).\gamma_k^{\pm} .$$

Enfin, on note

$$\gamma_{k,i}^{\pm} = \pi_i(\gamma_k^{\pm}) .$$

(On rappelle que les vecteurs $\gamma_k^{\pm}$, $k = 1,\ldots,6$ sont ceux déterminés par la prop. 4.7.1).

4.9 Détermination de L en fonction des périodes des courbes E_i.

4.9.1 Nous arrivons au point fondamental de ce paragraphe, dont le résultat final consistera à décrire le réseau L en termes des périodes des courbes E_i $(i = 1,\ldots,6)$.

Pour ce faire, nous allons utiliser l'action des opérateurs de Hecke sous la forme donnée par la proposition 3.3.1.

Soit $n=2$. La formule (3.3.1.1) s'écrit :

$$(4.9.1.1) \qquad \underline{c}(2).\{0,z\} = \{0,2z\} + \{0,\tfrac{z+1}{2}\} + \{0,\tfrac{z}{2}\} - \{0,\tfrac{1}{2}\} .$$

Utilisons les notations introduites en 4.8.2. La formule précédente est équivalente aux deux suivantes :

$$(4.9.1.2)^+ : \quad \underline{c}(2).\Big(\sum_{k=1}^{6} x_k^+(z).\gamma_k^+\Big) = \sum_{k=1}^{6} \Big[x_k^+(2z) + x_k^+(\tfrac{z}{2}) + x_k^+(\tfrac{z+1}{2}) - x_k^+(\tfrac{1}{2})\Big].\gamma_k^+ ,$$

$$(4.9.1.3)^- : \quad \underline{c}(2).\Big(\sum_{k=1}^{6} x_k^-(z).\gamma_k^-\Big) = \sum_{k=1}^{6} \Big[x_k^-(2z) + x_k^-(\tfrac{z}{2}) + x_k^-(\tfrac{z+1}{2})\Big].\gamma_k^-$$

soit encore, coordonnée par coordonnée, aux douze identités :

$$(4.9.1.2)_i^+ : \quad \sum_{k=1}^{6} \Big\{c_i(2).x_k^+(z) - \Big[x_k^+(2z) + x_k^+(\tfrac{z}{2}) + x_k^+(\tfrac{z+1}{2}) - x_k^+(\tfrac{1}{2})\Big]\Big\}.\gamma_{k,i}^+ = 0, \quad i=1,\ldots,6$$

$$(4.9.1.3)_i^- : \quad \sum_{k=1}^{6} \Big\{c_i(2).x_k^-(z) - \Big[x_k^-(2z) + x_k^-(\tfrac{z}{2}) + x_k^-(\tfrac{z+1}{2})\Big]\Big\}.\gamma_{k,i}^- = 0 . \quad i=1,\ldots,6$$

4.9.2 Ecrivons les formules $(4.9.1.2)_i^+$ pour les valeurs suivantes de z :

$$z = \frac{1}{3}, \frac{1}{36}, \frac{1}{8}, \frac{1}{51}, \frac{1}{26}, \frac{1}{18}$$

(rappelons que les six vecteurs $\{0,z\} + \{0,-z\}$ correspondants engendrent le réseau L^{+}).

On obtient ainsi un système linéaire homogène de six équations aux six inconnues $\gamma^{+}_{k,i}$, $1 \leq i,k \leq 6$. Les coefficients $x^{+}_{k}(z)$ se calculent de façon effective de la façon suivante :

(i) On exprime $\{0,\frac{b}{a}\}$ comme combinaison linéaire d'éléments contenus dans l'image de ξ , en utilisant la prop. 3.2.4.

(ii) Utilisant la table 1, on écrit les couples $(\tilde{c}:\tilde{d})$ en fonction des générateurs du groupe $\aleph(121)$.

(iii) Enfin, la table 3 permet d'écrire $\{0,\frac{b}{a}\}$ en fonction des générateurs γ^{+}_{k} et γ^{-}_{k} $(k=1,\ldots,6)$ des réseaux L^{+} et L^{-} .

Donnons à titre d'exemple le calcul de $x^{+}_{k}(\frac{2}{3})$, $k=1,\ldots,6$.

On a : $$\frac{2}{3} = 0 + \frac{1}{1+\frac{1}{2}} .$$

Par conséquent, avec les notations de la prop. 3.2.4 :

$$\frac{2}{3} = \frac{b_2}{a_2} , \quad \frac{1}{1} = \frac{b_1}{a_1} , \quad \frac{0}{1} = \frac{b_0}{a_0} .$$

On a donc :

$$\{0,\tfrac{2}{3}\} = \xi(-\tilde{1}:\tilde{1}) + \xi(\tilde{3}:\tilde{1}) .$$

Utilisant la table 1, on obtient :

$$(-\tilde{1}:\tilde{1}) = (1\tilde{2}0:\tilde{1}) = 0 .$$

Utilisant maintenant la table 3 :

$$\{0,\tfrac{2}{3}\} = \xi(\tilde{3}:\tilde{1}) = \{0,\tfrac{1}{3}\} = \tfrac{1}{2}(\gamma^{+}_{1} + \gamma^{-}_{1}) .$$

On en déduit, avec les notations de 4.8.2 :

$$x^{+}_{1}(\tfrac{2}{3}) = 1 \qquad x^{+}_{i}(\tfrac{2}{3}) = 0 \qquad 2 \leq i \leq 6 .$$

4.9.3 Si l'on exprime que le système linéaire homogène obtenu est de rang < 6 , on obtient :

$$c_i(2) \in \{0,\pm 1,\pm 2\} .$$

De façon plus précise, le système est de rang 5 pour

$c_i(2) = 0, \pm 1, +2$ et de rang 4 seulement lorsque $c_i(2) = -2$. (Si l'on se rappelle que la valeur propre -2 correspond aux formes f_1 , f_2 qui ne sont pas primitives, la difficulté rencontrée est prévisible). En tous cas, si l'on ne connaissait pas à l'avance les valeurs propres $c_i(2)$, $i = 1, \dots, 6$, on les obtiendrait à ce stade des calculs.

4.9.4 Considérons tout d'abord les cas $i = 3,4,5$ et 6 . On a donc $c_i(2) \neq -2$, et le système est de rang 5. On obtient comme solution générale, notant simplement $c_i(2) = c_i$:

$$\gamma^+_{1,i} = \frac{2c_i(c_i^2 + c_i - 1)}{c_i^2 - 2} \gamma^+_{4,i} ,$$

$$\gamma^+_{2,i} = c_i \gamma^+_{4,i} ,$$

$$\gamma^+_{3,i} = \frac{2c_i}{c_i^2 - 2} \gamma^+_{4,i} ,$$

$$\gamma^+_{5,i} = (1-c_i)\gamma^+_{4,i} ,$$

$$\gamma^+_{6,i} = (c_i^2-1)\gamma^+_{4,i} .$$

En particulier, ceci montre que les réseaux L_i^+ $(i = 3,4,5,6)$ sont engendrés par les nombres :

$$\omega_i^+ = \gamma^+_{4,i} \qquad (i = 3,4,5,6) .$$

Avec la numérotation choisie, on a (cf. prop. 1.4.1) :

$$c_3 = 2 , \ c_4 = 1 , \ c_5 = -1 , \ c_6 = 0 .$$

Portant ces valeurs dans la solution que l'on vient d'écrire, on obtient l'expression des i-ièmes coordonnées $(i = 3,4,5,6)$ des vecteurs γ_k^+ $(k = 1,\dots,6)$ en fonction de la période réelle ω_i^+ du réseau L_i^+ (cf. la table 4).

4.9.5 Passons maintenant aux cas $i = 1,2$ où $c_i = -2$.

a) Soit $i = 1$. On est ramené au calcul de symboles modulaires sur la courbe $X_o(11)$, qui est une courbe elliptique. On peut donc utiliser les résultats de Manin ([9], p. 59). Une base des réseaux $L_1^{\pm}$ est constituée par les vecteurs

$$\omega_1^{\pm} = \{o,\tfrac{1}{3}\}_{f_1} \pm \{o,-\tfrac{1}{3}\}_{f_1} .$$

Le réseau L_1 est alors le réseau d'indice 2 contenu dans $\frac{1}{2} L_1^+ \oplus \frac{1}{2} L_1^-$, qui contient lui-même $L_1 \oplus L_1^-$ comme sous-réseau d'indice 2.

L'expression des $\gamma_{k,1}^+$ $(k=1,\ldots,6)$ en fonction de ω_1^+ est donnée dans la table 4.

b) Soit $i=2$. On a $f_2(z) = f_1(z) - 11 f_1(11z)$ (cf. 1.2).

Il en résulte :

$$\{z_o,z_1\}_{f_2} = \{z_o,z_1\}_{f_1} - \{11z_o,11z_1\}_{f_1}$$

pour tous $z_o,z_1 \in \mathfrak{H}^*$.

On est donc ramené à un calcul du même type que celui de 4.9.2 pour calculer les $\gamma_{k,2}^+$ $(k=1,\ldots,6)$. Le résultat figure dans la table 4.

4.9.6 Nous ne donnerons pas de détails concernant la détermination des $\gamma_{k,i}^-$ $(1 \leqslant k,i \leqslant 6)$. Le calcul est tout à fait parallèle au précédent. On part des identités $(4.9.1.3)_i^-$, $i=1,\ldots,6$, que l'on écrit pour les valeurs de z :

$$z = \frac{1}{3}, \frac{1}{9}, \frac{1}{5}, \frac{1}{7}, \frac{1}{49}, \frac{1}{36} .$$

On obtient de nouveau un système linéaire homogène, qui donnerait une fois de plus les valeurs propres $c_i(2)$ si on ne les connaissait déjà. Ce système suffit à déterminer les $\gamma_{k,i}^-$ lorsque $i=3,4,5,6$; lorsque $i=1,2$, on raisonne comme précédemment.

Les résultats obtenus figurent dans la table 5.

4.9.7 <u>Les réseaux</u> L_i.

Revenons maintenant aux réseaux L_i $(i=1,\ldots,6)$. Soient :

$$\omega_1^+ = \gamma_{1,1}^+ ,$$

$$\omega_2^+ = 5\omega_1^+ ,$$

$$\omega_i^+ = \gamma_{4,i}^+ \qquad (i=3,4,5,6).$$

Soient également :

$$\omega_1^- = \gamma_{1,1}^- ,$$

$$\omega_2^- = \omega_1^- ,$$

$$\omega_3^- = -\frac{1}{2}\gamma_{5,3}^- ,$$

$$\omega_4^- = \gamma_{5,4}^- ,$$

$$\omega_5^- = \gamma_{5,5}^- ,$$

$$\omega_6^- = \frac{1}{2}\gamma_{5,6}^- .$$

Les résultats de la table 5 montrent que les nombres $\omega_i^{\pm}$, $i=1,\ldots,6$, engendrent $L_i^{\pm}$. Plus précisément, un calcul élémentaire montre le résultat suivant :

Proposition 4.9.7.1 : *Pour chaque* $i=1,\ldots,6$, *le réseau* L_i *est le sous-réseau d'indice* 2 *du réseau* $\frac{1}{2}L^+ \oplus \frac{1}{2}L^-$ *défini par* :

$$L_i = \left\{ \frac{a_i}{2}\omega_i^+ + \frac{b_i}{2}\omega_i^- \;\middle|\; \begin{array}{l} a_i, b_i \in \mathbb{Z} \\ a_i \equiv b_i \pmod 2 \end{array} \right\} .$$

Corollaire. *Le noyau de l'isogénie canonique* :

$$\prod_{i=1}^{6} \varphi_i : J_o(121) \to \prod_{i=1}^{6} E_i$$

est contenu dans le sous-groupe des points de $J_o(121)$ *annulés par la multiplication par* 24.

En d'autres termes :

$$\bigcap_{i=1}^{6} \pi_i^{-1}(L_i) \subseteq \frac{1}{24} L .$$

En effet, soit $z = \frac{1}{2}\sum_{k=1}^{6}(a_k\gamma_k^+ + b_k\gamma_k^-)$, $a_k, b_k \in \mathbb{R}$, un vecteur de $\mathbb{C}^6$.

Soient $\pi_i(z) = s_i w_i^+ + t_i w_i^-$, $i=1,\ldots,6$ les projections de z. Les expressions de a_k, b_k en fonction des s_i, t_i sont données par la table 6.

La condition pour que $z \in \bigcap_{i=1}^{6} \pi_i^{-1}(L_i)$ s'écrit :

(4.9.7.2) $$s_i \in \frac{1}{2}\mathbb{Z} ,\ t_i \in \frac{1}{2}\mathbb{Z} ,\ s_i + t_i \in \mathbb{Z} ,\quad i=1,\ldots,6 .$$

Soit $z' = 24z = \frac{1}{2}\sum_{k=1}^{6}(a'_k\gamma_k^+ + b'_k\gamma_k^-)$, donc : $a'_k = 24a_k$, $b'_k = 24b_k$. Remplaçons a'_k , b'_k par leur expression en fonction de s_k , t_k tirée de la table 6. On constate que les conditions (4.9.7.2) entraînent que a'_k et b'_k sont entiers et vérifient les conditions de la proposition 4.7.2. Par conséquent, $z' = 24z \in L$.

4.10 Détermination de l'ordre des pointes.

Soit $z \in \mathfrak{h}^*$ une pointe de $X_o(121)$. On se propose de déterminer l'ordre de cette pointe dans la jacobienne. Pour cela, on va déterminer $\{0,z\}$ en fonction des $\omega_i^{\pm}$, $i = 1,\ldots,6$.

4.10.1 Commençons par la détermination de $\{0,i\infty\}$ qui est plus facile.

Soit $z = i\infty$. Portant cette valeur dans (4.9.1.1), on obtient :

(4.10.1.1) $$(\underline{c}(2) - 3)\{0,i\infty\} = -\{0,\tfrac{1}{2}\} .$$

D'après la table 3 :

$$\{0,\tfrac{1}{2}\} = \gamma_1^+ - \gamma_2^+ - \gamma_3^+ - \gamma_4^+ - \gamma_6^+ .$$

Projetant les deux membres de (4.10.1.1) sur chaque coordonnée, et remplaçant les termes $\gamma_{k,i}^+$, $1 \leq k \leq i$ par leurs valeurs tirées de la table 4, on obtient :

$$\{0,i\infty\}_{f_1} = \frac{\omega_1^+}{5} , \quad \{0,i\infty\}_{f_2} = 0 ,$$
$$\{0,i\infty\}_{f_3} = 2\omega_3^+ , \quad \{0,i\infty\}_{f_4} = -\omega_4^+ ,$$
$$\{0,i\infty\}_{f_5} = -\omega_5^+ , \quad \{0,i\infty\}_{f_6} = 0 .$$

En particulier, ceci montre que l'image de la pointe O est d'ordre 5 dans la jacobienne $J_o(121)$. Plus précisément, son image sur E_1 est un point d'ordre 5, et ses images sur E_i , $i = 2,\ldots,6$ coïncident avec l'origine.

4.10.2 Ordre des pointes $z = \frac{u}{11}$, $u = 1,\ldots,10$.

Il est commode à ce point d'introduire une nouvelle notation. Soit u un entier rationnel. On pose :

(4.10.2.1) $$2\{i\infty,\frac{u}{11}\}_{f_i} = X(u,i)\omega_i^+ + Y(u,i)\omega_i^- \ , \quad 1 \leqslant i \leqslant 6 \ .$$

On sait a priori que les fonctions $X(u,i)$ resp. $Y(u,i)$ sont à valeurs dans $\mathbb{Q}$. De plus $X(u,i)$ resp. $Y(u,i)$ est une fonction paire resp. impaire de la variable u .

Ecrivons maintenant la formule 4.9.1.1 pour $z = \frac{u}{11}$. On obtient :

$$\underline{c}(2)\{0,\frac{u}{11}\} = \{0,\frac{2u}{11}\} + \{0,\frac{u+11}{22}\} + \{0,\frac{u}{22}\} - \{0,\frac{1}{2}\} \ .$$

Remplaçons u par $-u$, et ajoutons les deux formules obtenues. Faisons parcourir à u les valeurs $1,3,4,5,9$. Après projection sur chacune des coordonnées, on obtient un système de $5 \times 6 = 30$ équations linéaires aux inconnues $X(u,i)$, $u = 1,3,4,5,9$, $i = 1,\ldots,6$. La résolution de ce système donne les résultats qui figurent dans la table 7.

De façon analogue, on obtient un système de 30 équations linéaires aux inconnues $Y(u,i)$, $u = 1,3,4,5,9$, $i = 1,\ldots,6$, en retranchant les formules relatives à u et à $-u$.

Ce système se résoud et donne les résultats indiqués par la table 7.

4.11 Nous allons maintenant utiliser les identités :

$$f_{1,\chi_{11}} = f_3 \ , \ f_{4,\chi_{11}} = f_{5,\chi_{11}} = f_4 \ , \ f_{6,\chi_{11}} = f_6 \ ,$$

démontrées au §1. (Rappelons que χ_{11} désigne le caractère de Legendre mod 11).

4.11.1 Remarquons tout d'abord que si l'on pose :

$$f_o(z) = f_1(z) - f_1(11z) \ , \ \text{soit}$$

$$f_o = \frac{10f_1 + f_2}{11} \ ,$$

on a : $$f_{3,\chi_{11}} = f_o \ .$$

4.11.2 Remarquons également qu'on aurait pu remplacer la forme f_2 par la forme f_o dans le §4. On aurait obtenu un morphisme

$$\varphi_o : J_o(121) \to E_o \ .$$

Les tables 4 et 5 montrent que le réseau L_o définissant E_o est engendré par les vecteurs $\frac{1}{2}\,\omega_o^+$, $\frac{1}{2}(\omega_o^+ + \omega_o^-)$, où

$$\omega_o^+ = \frac{5}{11}\,\omega_1^+ \ , \ \omega_o^- = \frac{1}{11}\,\omega_1^- \ .$$

Enfin,

$$\{i\infty,0\}_{f_o} = \frac{1}{11}\Big[10\{i\infty,0\}_{f_1} + \{i\infty,0\}_{f_2}\Big] = -\frac{2\omega_1^+}{11} = -\frac{2}{5}\,\omega_o^+ \ (\text{cf.}4.10.1).$$

Par conséquent, si l'on étend la notation (4.10.2.1) au cas où $f_i = f_o$, on a :

$$X(0,0) = -\frac{4}{5} \ , \ Y(0,0) = 0 \ .$$

4.11.3 Rappelons maintenant le résultat suivant (cf [9], 3.9) :

<u>Lemme</u> 4.11.3.1 : <u>Soit</u> $f \in \langle\Gamma_o(121),2\rangle_o$, <u>et</u> χ <u>un caractère de Dirichlet primitif modulo</u> 11. <u>Soit</u> f_χ <u>la forme tordue de</u> f <u>par</u> χ . <u>On a alors</u> :

$$\{0,z\}_{f_\chi} = -\frac{i}{\sqrt{11}}\sum_{u=1}^{10} \chi(u)\ \{\tfrac{u}{11}\ ,\ z+\tfrac{u}{11}\}_f \ .$$

Nous allons appliquer ce résultat aux formes f_i ; soit donc $i = 0,1,\ldots,6$, et définissons i' par :

$$f_{i,\chi_{11}} = f_{i'} \ .$$

Considérons d'abord le cas où $z = i\infty$. On obtient l'identité :

$$\{i\infty,0\}_{f_{i'}} = \frac{-i}{\sqrt{11}}\sum_{u=1}^{10} \chi_{11}(u)\ \{i\infty,\tfrac{u}{11}\}_{f_i} \ ,$$ soit encore,

utilisant les notations (4.10.2.1) :

(4.11.3.2) $$X(0,i')\,\omega_{i'}^+ = -\frac{2i}{\sqrt{11}}\sum_{\substack{u \bmod 11\\ \chi_{11}(u)=1}} Y(u,i).\omega_i^- \ .$$

4.11.4 Appliquons la formule précédente dans les cas suivants :

(a) $i=3$, $i'=0$; tenant compte de ce que :

$$X(0,0) = -\frac{4}{5} \ \text{ et } \ \omega_o^+ = \frac{5}{11}\,\omega_1^+ \ ,$$ on obtient :

$$\omega_3^- = -\frac{5i}{\sqrt{11}}\,\omega_1^+ \ .$$

(b) $i=1$, $i'=3$; avec l'aide de la table 7, on trouve :

$$\omega_3^+ = \frac{i}{\sqrt{11}}\,\omega_1^- \ .$$

(c) $i=4$, $i'=5$ et $i=5$, $i'=4$, donnent :

$$\omega_5^+ = -\frac{i}{\sqrt{11}}\,\omega_4^- \quad \text{et}$$

$$\omega_5^- = i\sqrt{11}\,\omega_4^+ \,.$$

4.11.5 Pour $i=6$, $i'=6$, les deux membres de (4.11.3.2) sont nuls. Il faut donc utiliser un point autre que $z = i\infty$, par exemple $z = \frac{1}{11}$ pour obtenir une identité non triviale.

Faisant $z = \frac{1}{11}$, $f = f_6$ dans le lemme 4.11.3.1, on obtient :

$$\omega_6^- = \frac{i}{\sqrt{11}}\,\omega_6^+ \,.$$

5. <u>Interprétation des résultats</u>.

5.1 Nous utiliserons à plusieurs reprises le résultat suivant :

<u>Lemme</u> 5.1.1 : <u>Soient</u> X_1 , X_2 <u>deux courbes elliptiques sur</u> $\mathbb{Q}$. <u>On suppose que</u> X_1 <u>et</u> X_2 <u>deviennent isomorphes sur</u> $\mathbb{C}$, <u>et ont bonne réduction en dehors de</u> 11.

<u>Alors, ou bien</u> X_1 <u>est isomorphe à</u> X_2 <u>sur</u> $\mathbb{Q}$, <u>ou bien</u> X_1 <u>est tordue de</u> X_2 <u>sur</u> $\mathbb{Q}(\sqrt{-11})$.

Tout d'abord, X_1 et X_2 n'ont pas d'autres automorphismes que ± 1 . En effet, une courbe elliptique d'invariant $j = 1728$, resp. $j=0$, a nécessairement mauvaise réduction en 2 , resp. en 3 . Par conséquent, X_1 et X_2 sont isomorphes sur une extension de $\mathbb{Q}$ de degré au plus 2. De plus, cette extension est non ramifiée en dehors de 11. C'est donc ou bien $\mathbb{Q}$, ou bien $\mathbb{Q}(\sqrt{-11})$.

5.2 <u>Les courbes</u> E_1 , E_2 .

5.2.1 Considérons la courbe E_1 . Cette courbe correspond au réseau L_1 engendré par les vecteurs ω_1^+ et $\frac{1}{2}(\omega_1^+ + \omega_1^-)$ dans $\mathbb{C}$. Or ce réseau est défini uniquement en termes de la forme f_1 . Plus précisément, la prop. 3.1.2 appliquée à f_1 , base de $\langle\Gamma_0(11),2\rangle_0$, montre que $X_0(11) = J_0(11)$ est définie, en tant que courbe sur $\mathbb{C}$, par le réseau L_1 .

Utilisons le lemme 5.1.1. Comme les coefficients de la série L de E_1 et $X_o(11)$ coïncident, et que $f_{1,X_{11}} \neq f_1$, les courbes E_1 et $X_o(11)$ ne sont pas tordues l'une de l'autre. Elles sont donc isomorphes sur $\mathbb{Q}$. L'image de la pointe 0 engendre dans E_1 un sous-groupe cyclique d'ordre 5, noté C_1 . Ce groupe correspond au sous-groupe engendré dans L_1 par $\frac{1}{5}\omega_1^+$.

5.2.2 La courbe E_2 correspond au réseau L_2 engendré par les vecteurs ω_2^+ et $\frac{1}{2}(\omega_2^+ + \omega_2^-)$, avec $\omega_2^+ = 5\omega_1^+$ et $\omega_2^- = \omega_1^-$. La pointe 0 a pour image l'origine de E_2 , et les autres pointes engendrent le groupe C_2 défini par le vecteur $\frac{1}{25}\omega_2^+$. Considérons le quotient $E_2/5C_2$. Il s'agit de la courbe elliptique correspondant au réseau défini par $\frac{1}{5}\omega_2^+$ et $\frac{1}{2}(\omega_2^+ + \omega_2^-)$, qui n'est autre que le réseau L_1 .

On a donc une suite d'isogénies de degré 5 :

$$\begin{array}{ccccc} E_2 & \to & E_1 & \to & E_2/C_2 \\ & & \wr\| & & \wr\| \\ & & E_2/5C_2 & & E_1/C_1 \end{array} .$$

5.2.3 Examinons l'action du groupe de Galois sur les pointes. D'après Ogg [11] (cf. également II, §3), les pointes de $X_o(121)$ sont rationnelles sur le corps cyclotomique $\mathbb{Q}(\zeta_{11})$; de plus, si l'on désigne par $P_{11,u}$ la pointe correspondant à $z = \frac{u}{11}$ $(u = 1,\ldots,10)$, l'élément $[n]$ de $\mathrm{Gal}(\mathbb{Q}(\zeta_{11})/\mathbb{Q})$ agit par :

$$P_{11,u}^{[n]} = P_{11,n^{-1}u} ,$$

la pointe 0 étant elle-même rationnelle sur $\mathbb{Q}$.

Par conséquent, le groupe C_1 est rationnel sur $\mathbb{Q}$ point par point.

D'autre part, la table 7 montre que l'on a, avec les notations de la prop. 2.6.1 :

$$\varphi_2(P_{11,u}) = \varepsilon(u^4).\varphi_2(P_{11,1}) .$$

Le morphisme φ_2 étant $\mathbb{Q}$-rationnel, on en déduit :

$$\varphi_2(P_{11,u})^{[n]} = \varphi_2(P_{11,n^{-1}.u}) = \frac{\varepsilon(u^4)}{\varepsilon(n^4)}\varphi_2(P_{11,1})$$
$$= \varepsilon(n)^6\varphi_2(P_{11,u}) = \chi_{11}(n)\varepsilon(n)\varphi_2(P_{11,u}) .$$

C'est la structure galoisienne de C_2 annoncée dans la prop. 2.6.1.

5.2.4 En particulier, les groupes C_2 , $5C_2$ sont globalement rationnels sur $\mathbb{Q}$. Les isogénies correspondantes sont donc définies sur $\mathbb{Q}$.

De plus, il résulte de :

$$\chi_{11}(n)\varepsilon(n) \equiv 1 \pmod 5$$

que tous les points du groupe $5C_2$ sont rationnels sur $\mathbb{Q}$.

En résumé, le diagramme :

$$E_2 \to E_1 \to E_1/C_1 \qquad \text{(cf. 5.2.2)}$$

est formé d'isogénies $\mathbb{Q}$-rationnelles de degré 5, et les courbes E_1 , E_2 contiennent chacune un point d'ordre 5 rationnel sur $\mathbb{Q}$.

5.2.5 On a déjà vu que E_1 est la courbe $X_o(11)$. La courbe E_1/C_1 est le quotient de $X_o(11)$ par le sous-groupe d'ordre 5 engendré par les pointes. Par conséquent, la courbe E_2 est la troisième courbe isogène sur $\mathbb{Q}$ à $X_o(11)$ (cf. [7], 7.5.1) : c'est la courbe $X_1(11)$, notée 11A dans la table 1 de [2]. Les courbes E_1 , E_1/C_1 sont donc les courbes 11B , 11C . (Dans [7], les courbes E_2 , E_1 , E_1/C_1 sont notées A_{11} , X_{11} , C_{11} resp.).

5.3 <u>Les courbes</u> E_3 , $E_3/5C_3$, E_3/C_3 .

5.3.1 La courbe E_3 correspond au réseau L_3 engendré par ω_3^+ et $\frac{1}{2}(\omega_3^+ + \omega_3^-)$. D'après 4.11.4, a, b, ce réseau n'est autre que le réseau $\frac{i}{\sqrt{11}} L_2$. Les courbes E_2 , E_3 sont donc isomorphes sur $\mathbb{C}$.

Le lemme 5.1.1, joint au fait que les séries L de ces deux courbes sont distinctes, entraîne que E_2 et E_3 sont tordues l'une de l'autre sur $\mathbb{Q}(\sqrt{-11})$: $E_3 \cong E_2'$.

5.3.2 La table 7 montre que l'on a :

$$\varphi_3(P_{11,u}) = \varepsilon(u^9).\varphi_3(P_{11,1}) .$$

On en déduit :

$$\varphi_3(P_{11,u})^{[n]} = \varepsilon(n).\varphi_3(P_{11,u}) .$$

C'est bien là l'action galoisienne annoncée par la prop. 2.6.1.

Du reste, ceci résulte également du fait que E_3 est tordue de E_2 sur $\mathbb{Q}(\sqrt{-11})$, et de l'action connue du groupe de Galois sur C_2.

5.3.3 On obtient, de façon analogue au cas de E_2, deux isogénies de degré 5, définies sur $\mathbb{Q}$:

$$E_3 \to E_3/5C_3 \to E_3/C_3 .$$

On peut aussi considérer que ce diagramme se déduit par torsion du diagramme

$$E_2 \to E_1 \to E_1/C_1 .$$

5.3.4 Montrons comment l'utilisation des endomorphismes de torsion définis en 3.4 permet de retrouver le fait que les courbes E_2 et E_3 sont tordues l'une de l'autre sur $\mathbb{Q}(\sqrt{-11})$, et ceci sans avoir à utiliser le lemme 5.1.1.

Considérons l'endomorphisme R_{11} de $J = J_o(121)$. On montre facilement (cf. [13], prop. 8) l'existence d'isogénies ν_o, ν_3, définies sur $\mathbb{Q}(\sqrt{-11})$, et telles que le diagramme suivant soit commutatif :

$$\begin{array}{ccccc} J & \xrightarrow{R_{11}} & J & \xrightarrow{R_{11}} & J \\ \downarrow{\scriptstyle\varphi_o} & & \downarrow{\scriptstyle\varphi_3} & & \downarrow{\scriptstyle\varphi_o} \\ E_o & \xrightarrow{\nu_o} & E_3 & \xrightarrow{\nu_3} & E_o \end{array}$$

(on rappelle que φ_o désigne le morphisme associé à la forme f_o définie en 4.11.1).

On a déjà remarqué que la courbe E_o est définie par le réseau L_o engendré par $\frac{5}{11}\omega_1^+$ et $\frac{1}{2}(\frac{5}{11}\omega_1^+ + \frac{1}{11}\omega_1^-)$. Par conséquent, $L_o = \frac{1}{11}L_2$, et il est facile d'en déduire que $E_o \cong E_2$.

Sur les formes paraboliques, R_{11} induit l'opérateur R_{11}^{*} (cf. 3.4), et l'on a : $f_o|R_{11}^{*} = i\sqrt{11}\ f_3$, $f_3|R_{11}^{*} = i\sqrt{11}\ f_o$. Il en résulte que $\nu_3 \circ \nu_o$ est la multiplication par -11. Comme E_o , E_3 n'ont pas de multiplication complexe, ceci entraîne que l'une des isogénies est un isomorphisme. On vérifie facilement, en interprétant R_{11} en termes de réseaux, que ν_o est un isomorphisme, et que ν_{11} a pour noyau le noyau de la multiplication par 11.

5.4 <u>Les courbes</u> E_4 , E_5 , E_4/C_4 , E_5/C_5 .

5.4.1 La courbe E_4 correspond au réseau L_4 engendré par ω_4^{+} et $\frac{1}{2}(\omega_4^{+}+\omega_4^{-})$. La table 7 montre que le groupe cuspidal C_4 est engendré par $\frac{1}{11}\omega_4^{-}$, donc cyclique d'ordre 11, et que de plus :

$$\varphi_4(P_{11},u) = u^3.\varphi_4(P_{11,1}).$$

Utilisant :

$$\varphi_4(P_{11,u})^{[n]} = \varphi_4(P_{11,n^{-1}u}),$$

on en déduit :

$$\varphi_4(P_{11,u})^{[n]} = n^7.\varphi_4(P_{11,1}),$$

ce qui montre que le module galoisien C_4 est isomorphe à $\mu_{11}^{\otimes 7}$.

5.4.2 La courbe E_5 correspond au réseau L_5 engendré par ω_5^{+} et $\frac{1}{2}(\omega_5^{+}+\omega_5^{-})$. D'après 4.11.4, c, ce réseau est aussi celui engendré par $-\frac{i}{\sqrt{11}}\ \omega_4^{-}$ et $-\frac{i}{\sqrt{11}}\ (\omega_4^{-}-11\omega_4^{+})$. On reconnait là un réseau homothétique au réseau qui définit E_4/C_4 . Comme E_5 et E_4/C_4 ne sont pas isomorphes sur $\mathbb{Q}$ (leurs séries L sont distinctes), on a nécessairement :

$$E_5 \cong (E_4/C_4)' .$$

5.4.3 L'existence de la forme bilinéaire de Weil sur les points de E_4 annulés par 11 entraîne l'existence dans E_4/C_4 d'un sous-groupe isomorphe à $\mu_{11}^{\otimes 4}$ comme module galoisien. On en déduit par torsion l'existence dans E_5 d'un groupe isomorphe à $\mu_{11}^{\otimes 9}$. Ce résultat se retrouve directement comme suit : la table 7 montre que le groupe cuspidal C_5 est engendré par $\frac{1}{11}\omega_5^{-}$, donc cyclique d'ordre 11, et que

$$\varphi_5(P_{11,u}) = u.\varphi_5(P_{11,1}), \qquad u = 1,\dots,10 .$$

On en déduit :

$$\varphi_5(P_{11,u})^{[n]} = n^9.\varphi_5(P_{11,u}),$$ ce qui montre que C_5 est isomorphe à $\mu_{11}^{\otimes 9}$.

On montre comme plus haut :

$$E_4 \cong (E_5/C_5)' .$$

5.4.4 On peut là encore retrouver une partie de ces résultats en utilisant les endomorphismes de torsion, ce qui évite l'emploi du lemme 5.1.1.

En effet, on peut montrer l'existence d'isogénies ν_4, ν_5 définies sur $\mathbb{Q}(\sqrt{-11})$, rendant commutatif le diagramme :

$$\begin{array}{ccccc} J & \xrightarrow{R_{11}} & J & \xrightarrow{R_{11}} & J \\ \varphi_4\downarrow & & \varphi_5\downarrow & & \downarrow\varphi_4 \\ E_4 & \xrightarrow{\nu_4} & E_5 & \xrightarrow{\nu_5} & E_4 \end{array} .$$

Là encore, on déduit de

$$f_4|R_{11}^* = i\sqrt{11}\, f_5 \quad , \quad f_5|R_{11}^* = i\sqrt{11}\, f_4 ,$$

que $\nu_5 \circ \nu_4$ est la multiplication par -11. Nous allons voir qu'en fait :

$$\deg \nu_4 = \deg \nu_5 = 11$$

en exhibant un point d'ordre 11 qui se trouve dans le noyau de ν_4, resp. ν_5

En effet, considérons la pointe $P_{11,1}$. On a déjà remarqué que sa projection $\varphi_4(P_{11,1})$ resp. $\varphi_5(P_{11,1})$ engendre C_4 resp. C_5.

D'une part, notant P_o la pointe correspondant à $z=0$, on a :

(5.4.4.1) $$R_{11}(P_{11,1}) = P_{11,2} + P_{11,4} + P_{11,5} + P_{11,6} + P_{11,10} - P_{11,3} - P_{11,7} - P_{11,8} - P_{11,9} - P_o ,$$

par définition même de R_{11}.

Or $\varphi_4(P_o) = 0$, $\varphi_5(P_o) = 0$, et

$$\varphi_4(P_{11,u}) = u^3\ \varphi_4(P_{11,1})\ ,$$

$$\text{resp.}\quad \varphi_5(P_{11},u) = u\ \ \varphi_5(P_{11,1})\ .$$

On en déduit :

$$\varphi_4 \circ R_{11}(P_{11,1}) = 0\ ,\ \varphi_5 \circ R_{11}(P_{11,1}) = 0\ ,$$

ce qui montre que le noyau de ν_4 resp. ν_5 est le groupe cuspidal C_4 resp. C_5 .

Comme les isogénies ν_4 , ν_5 ne sont pas définies sur $\mathbb{Q}$, ce sont nécessairement les composées de l'isogénie naturelle :

$$E_4 \to E_4/C_4\ ,$$

$$\text{resp.}\quad E_5 \to E_5/C_5\ ,$$

avec la torsion sur $\mathbb{Q}(\sqrt{-11})$.

5.5 <u>La courbe</u> E_6 .

5.5.1 La courbe E_6 est définie par le réseau L_6 , engendré par ω_6^- et $\frac{1}{2}(\omega_6^+ + \omega_6^-)$. D'après 4.11.5, ce réseau est engendré par ω_6^- et $\omega_6^-\ (\frac{1 - i\sqrt{11}}{2})$. On a donc :

$$L_6 = \omega_6^-.O_K$$

où O_K désigne l'anneau des entiers du corps $K = \mathbb{Q}(\sqrt{-11})$.

5.5.2 Le groupe cuspidal C_6 correspond au groupe engendré par le vecteur $\frac{1}{11}\omega_6^+ = -\frac{i}{\sqrt{11}}\omega_6^-$. Il est donc cyclique d'ordre 11. De plus :

$$\varphi_6(P_{11,u}) = u^2.\varphi_6(P_{11,1})\ ,\quad u = 1,\dots,10\ .$$

On en déduit :

$$\varphi_6(P_{11,u})^{[n]} = n^8.\varphi_6(P_{11,u})\ .$$

5.5.3 La courbe E_6 admet une multiplication complexe par l'anneau O_K . On vérifie immédiatement que C_6 est le noyau de l'endomorphisme de multiplication par $-i\sqrt{11}$. En particulier, ceci entraîne que E_6/C_6 est isomorphe à la courbe E_6' , tordue de E_6 sur $\mathbb{Q}(\sqrt{-11})$.

5.5.4 On peut ici également considérer l'endomorphisme de torsion R_{11}. On montre l'existence d'une isogénie ν_6, définie sur $\mathbb{Q}(\sqrt{11})$, telle que le diagramme

$$\begin{array}{ccc} J & \xrightarrow{R_{11}} & J \\ \varphi_6 \downarrow & & \downarrow \varphi_6 \\ E_6 & \xrightarrow{\nu_6} & E_6 \end{array}$$

soit commutatif. L'identité $f_6 | R_{11}^* = i\sqrt{11}\ f_6$ entraîne que $\nu_6^2 = -11$.

On peut alors utiliser l'expression de $R_{11}(P_{11,1})$ donnée en (5.4.4.1). Du fait que

$$\varphi_6(P_{11,u}) = u^2.\varphi_6(P_{11,1})$$

résulte que :

$$\varphi_6 \circ R_{11}(P_{11,1}) = \nu_6(\varphi_6(P_{11,1})) = 0 .$$

Par conséquent, le noyau de ν_6 est le groupe C_6. On en déduit

$$E_6/C_6 \cong E_6' .$$

5.6 <u>Comparaison avec les courbes de Swinnerton-Dyer et Vélu</u>.

On peut maintenant énoncer :

<u>Proposition</u> 5.6.1 : <u>Soit</u> E <u>une courbe elliptique définie sur</u> $\mathbb{Q}$, <u>ayant bonne réduction en dehors de</u> 11. <u>On suppose que</u> E <u>contient un groupe cyclique</u> C <u>d'ordre</u> 11, <u>rationnel sur</u> $\mathbb{Q}$.

<u>Alors</u> E <u>est isomorphe à l'une des six courbes</u> E_i, E_i', $i = 4,5,6$, <u>et le groupe</u> C <u>est isomorphe comme module galoisien au groupe</u> $\mu_{11}^{\otimes a}$. <u>Plus précisément</u>

$a = 7,9,8$, resp. $2,4,3$ <u>selon que</u> $E = E_i$, $i = 4,5,6$ resp. $E = E_i'$, $i = 4,5,6$.

Remarquons tout d'abord que les trois couples $(E_4, E_4/C_4)$, $(E_5, E_5/C_5)$, $(E_6, E_6/C_6)$ constituent trois solutions $\mathbb{Q}$-rationnelles du problème modulaire associé à la courbe $X_o(11)$. Plus précisément, on sait qu'à un couple (E,C) vérifiant les conditions de la proposition est associé un point de $X_o(11)$, rationnel sur $\mathbb{Q}$, et distinct des pointes de $X_o(11)$. D'autre part, les invariants $j(E_4)$, $j(E_5)$ et $j(E_6)$ sont distincts, par exemple parce que E_6 admet une multiplication complexe, ce qui n'est pas le cas de E_4 et E_5. Ceci entraîne que les trois points correspondants de $X_o(11)$ sont distincts.

On sait aussi que la courbe $X_o(11)$ n'a que trois points rationnels sur $\mathbb{Q}$ autres que les pointes. Le couple $(E,E/C)$ coïncide donc, en tant que solution du problème modulaire, avec l'un des trois précédents. En particulier, il existe un i, $1 \leqslant i \leqslant 6$ tel que $j(E) = j(E_i)$. On peut donc appliquer le lemme 5.1.1 aux courbes E, E_i, et la première conclusion de la proposition en résulte.

D'autre part, il est exclu que tous les points d'ordre 11 de E soient rationnels sur le corps cyclotomique $\mathbb{Q}(\zeta_{11})$: Si cela était le cas, la réduction de E modulo ℓ, pour tout ℓ premier tel que $\ell \equiv 1 \pmod{11}$, par exemple $\ell = 23$, aurait un nombre de points rationnels sur $\mathbb{F}_\ell$ congru à $0 \pmod{121}$; or ce n'est pas le cas.

Par conséquent, la courbe E contient un $\mu_{11}^{\otimes a}$, avec un a bien déterminé qui caractérise la courbe en question.

Remarquons ici que les valeurs $a = 0,1,5,6 \pmod{10}$ sont exclues : La valeur 0 parce que la courbe $X_1(11)$ n'a pas d'autres points rationnels que les pointes. Les autres, parce qu'on se ramènerait à $a=0$ en tordant la courbe, ou en considérant le quotient par le groupe en question.

5.6.2 Considérons les courbes notées 121D,E,F,G,H,I dans [2]. Elles vérifient toutes les hypothèses de la prop. 5.6.1. Ce sont donc, à l'ordre près les courbes E_i, E_i', $i = 4,5,6$. Si l'on tient compte des renseignements sur les séries L d'une part, et sur le fait que l'on connait laquelle d'entre ces courbes est une courbe de Weil, on arrive à la correspondance donnée par la prop. 2.6.1. Les invariants correspondants sont tirés de la liste de Vélu [16].

Enfin, toute courbe elliptique définie sur $\mathbb{Q}$, isogène sur $\mathbb{Q}$ à l'une des douze courbes E_i , E_i' , $i = 1,\dots,6$, E_1/C_1 , E_3/C_3 (rappelons que $E_2' = E_3$) coïncide avec l'une d'entre elles : En effet, les courbes de la liste de Vélu [16] possèdent cette propriété. On a donc déterminé toutes les courbes elliptiques quotients sur $\mathbb{Q}$ de $J_o(121)$.

II. Courbes associées aux sous-groupes de $SL_2(\mathbb{F}_{11})$

Avant d'aborder, au §4, l'étude des courbes associées aux sous-groupes de $SL_2(\mathbb{F}_{11})$, nous allons rappeler un certain nombre de résultats généraux, concernant les courbes associées à des sous-groupes de $SL_2(\mathbb{F}_p)$, pour p premier, $p \geqslant 5$. La démonstration des affirmations non justifiées des §2 et 3 se trouve dans [8].

1. Notations. Toutes les notations que nous allons introduire ont cours dans l'ensemble du chapitre II. Simplement, on supposera que $p = 11$ à partir du §4.

Soit p un nombre premier $\geqslant 5$.

Soit H un sous-groupe de $SL_2(\mathbb{F}_p)$, contenant la matrice -1. On notera $\tilde{H}$ l'image inverse de H dans $SL_2(\mathbb{Z})$ par le morphisme de réduction modulo p . Le groupe $\tilde{H}$ est un groupe de congruence de niveau p , c'est-à-dire contenant $\Gamma(p)$.

On notera X_H la courbe algébrique associée à $\tilde{H}$. En d'autres termes, les points de X_H à valeurs dans $\mathbb{C}$ s'identifient avec la surface de Riemann compacte quotient du demi-plan de Poincaré complété $\mathfrak{H}^*$ par l'action de $\tilde{H}$. En particulier, l'injection naturelle de $\Gamma(p)$ dans $\tilde{H}$ définit un morphisme canonique

$$X(p) \to X_H$$

qui permet aussi de considérer X_H comme le quotient de $X(p)$ par le sous-groupe d'automorphismes défini par H .

Une conséquence de ceci est que les pointes de X_H s'identifient aux orbites de l'action de H sur les pointes de $X(p)$.

2. <u>Les courbes</u> X_H <u>et leurs pointes</u> : <u>Rappels</u>.

2.1 <u>Genre de</u> X_H (cf. [8]).

2.1.1 Le genre de la courbe X_H est donné par la formule :

$$\text{genre de } X_H = 1 + \frac{\mu_H}{12} - \frac{e_{2,H}}{4} - \frac{e_{3,H}}{3} - \frac{e_{\infty,H}}{2},$$

où :

$$\mu_H = [SL_2(\mathbb{Z}):\widetilde{H}] = [SL_2(\mathbb{F}_p):H] = \frac{p(p^2-1)}{|H|} ;$$

$e_{2,H}$ est le nombre de points elliptiques de $\widetilde{H}$ d'ordre 2 ou 4 ;

$e_{3,H}$ est le nombre de points elliptiques de $\widetilde{H}$ d'ordre 3 ou 6 ;

$e_{\infty,H}$ est le nombre de pointes de $\widetilde{H}$.

2.1.2 <u>Calcul de</u> $e_{\infty,H}$.

a) Supposons que l'ordre de H soit premier à p . Alors $\bar{H} = H/\{\pm 1\}$ opère librement sur les pointes de $X(p)$, d'où :

$$e_{\infty,H} = \frac{p^2-1}{|H|} = \frac{\mu_H}{p} .$$

b) Soit δ un diviseur positif de $\frac{p-1}{2}$, et posons $p-1 = 2\delta\delta'$. Prenons pour H le groupe $H_{1,\delta}$ défini de la façon suivante :

$$H_{1,\delta} = \left\{\pm\begin{pmatrix} a & b \\ 0 & a^{-1}\end{pmatrix} \;\middle|\; \begin{matrix} b \in \mathbb{F}_p \\ a \in \mathbb{E}_p^{\delta}\end{matrix}\right\} ,$$

où l'on note $\mathbb{E}_p = \mathbb{F}_p^{\times}/\{\pm 1\}$, et $\mathbb{E}_p^{\delta}$ = le sous-groupe de $\mathbb{E}_p$ formé des éléments d'ordre divisant δ'.

(Le groupe $H_{1,\delta}$ est d'ordre $2p\delta'$, d'où $\mu_H = (p+1)\delta$).

On a alors :

$$e_{\infty,H} = 2\delta .$$

2.1.3 <u>Calcul de</u> $e_{2,H}$, $e_{3,H}$.

Les nombres $e_{2,H}$, $e_{3,H}$ sont donnés par la proposition suivante (cf. [8]) :

<u>Proposition</u> 2.1.3.1 : <u>Avec les notations précédentes</u>, <u>on a</u>

$$e_{2,H} = \frac{p-\left(\frac{-1}{p}\right)}{|H|}\,\mathrm{Card}\,\{h \in H \mid \mathrm{tr}(h) = 0\} ,$$

$$e_{3,H} = \frac{p - (\frac{-3}{p})}{|H|} \text{Card}\{h \in H \mid \text{tr } h = -1\} .$$

Exemples : a) Supposons que H soit un groupe de Cartan déployé, resp. non déployé, resp. le normalisateur de l'un des précédents. Les nombres $e_{2,H}$, $e_{3,H}$ sont donnés par le tableau suivant :

H	$e_{2,H}$	$e_{3,H}$
Cartan déployé	0 si $p \equiv 3 \pmod 4$ 2 si $p \equiv 1 \pmod 4$	0 si $p \equiv 2 \pmod 3$ 2 si $p \equiv 1 \pmod 3$
normalisateur de Cartan déployé	$\frac{p+1}{2}$	0 si $p \equiv 2 \pmod 3$ 1 si $p \equiv 1 \pmod 3$
Cartan non déployé	2 si $p \equiv 3 \pmod 4$ 0 si $p \equiv 1 \pmod 4$	0 si $p \equiv 1 \pmod 3$ 2 si $p \equiv 2 \pmod 3$
normalisateur de Cartan non déployé	$\frac{p+3}{2}$ si $p \equiv 3 \pmod 4$ $\frac{p-1}{2}$ si $p \equiv 1 \pmod 4$	0 si $p \equiv 1 \pmod 3$ 1 si $p \equiv 2 \pmod 3$

(Une partie des résultats de ce tableau figure dans [3]).

b) Supposons que $\bar{H} = H/\{\pm 1\}$ soit isomorphe au groupe alterné $\mathfrak{A}_4$. Ce groupe contient trois éléments d'ordre 2 et huit éléments d'ordre 3. Par conséquent, H contient 6 éléments de trace nulle et 8 de trace -1. Comme $|H| = 24$, on a :

$$e_{2,H} = \frac{p - (\frac{-1}{p})}{4}$$

et

$$e_{3,H} = \frac{p - (\frac{-3}{p})}{3} .$$

c) Supposons que H soit le groupe $H_{1,\delta}$ défini en 2.1.2, b. Un calcul facile montre que

$$e_{2,H} = \begin{cases} 2\delta & \text{si } p \equiv 1 \pmod 4 \text{ et } \delta' \equiv 0 \pmod 2, \\ 0 & \text{sinon ;} \end{cases}$$

$$e_{3,H} = \begin{cases} 2\delta & \text{si } p \equiv 1 \pmod 3 \text{ et } \delta' \equiv 0 \pmod 3, \\ 0 & \text{sinon.} \end{cases}$$

Par exemple, si $\delta = 1$, la courbe associée à $H_{1,1}$ n'est autre que la courbe $X_o(p)$. Si $\delta = \frac{p-1}{2}$, on a affaire à la courbe $X_1(p)$. Pour la première, on trouve

$$e_{2,H} = 1 + (\tfrac{-1}{p}) ,$$

$$e_{3,H} = 1 + (\tfrac{-3}{p}) .$$

Pour la seconde

$$e_{2,H} = e_{3,H} = 0 .$$

2.2 <u>Pointes de</u> X_H .

2.2.1 Sur l'ensemble de couples d'entiers $\tilde{\Pi}$:

$$\tilde{\Pi} = \left\{ \binom{u}{v} \mid u \in \mathbb{Z} \; ; \; v \in \mathbb{Z} \; ; \; (u,v) = 1 \right\}$$

définissons la relation d'équivalence :

$\binom{u'}{v'}$ est équivalent à $\binom{u}{v}$ si et seulement si :

$$\binom{u'}{v'} \equiv \pm \binom{u}{v} \pmod{p}.$$

Soit Π le quotient de $\tilde{\Pi}$ par cette relation d'équivalence. C'est un ensemble fini qui contient $\frac{p^2-1}{2}$ éléments.

L'ensemble Π s'identifie à l'ensemble des pointes de $X(p)$ de la façon suivante :

Soit $\binom{u}{v} \in \tilde{\Pi}$ un représentant de $\binom{u}{v} \in \Pi$, avec $v \neq 0$. On lui associe l'image de $\frac{u}{v} \in \mathbb{Q}$ par l'application canonique :

$$\mathfrak{H}^* = \mathbb{Q} \cup \{i\infty\} \cup \mathfrak{H} \rightarrow \Gamma(p) \backslash \mathfrak{H}^* = X(p)$$

qui est une pointe de $X(p)$ (cf. [11]).

On utilisera constamment cette identification par la suite.

Le groupe H opère à gauche sur Π , et l'ensemble des pointes de X_H s'identifie à l'ensemble $H \backslash \Pi$ des orbites de cette action.

2.2.2 Considérons une pointe de X_H , représentée par $\binom{u}{v} \in \Pi$. Soit $M \in SL_2(\mathbb{Z})$ une matrice telle que $M.\binom{u}{v} = \binom{1}{0}$. Il existe un plus petit entier $n > 0$ tel que la matrice :

$$M^{-1}.\begin{pmatrix} 1 & n \\ 0 & 1 \end{pmatrix} M$$

soit un élément de $\tilde{H}$ (en fait, $n = 1$ ou p).

2.3 Formes cuspidales.

2.3.1 Utilisons les notations et les résultats de [5].

Etant donné un couple (r,w) d'entiers, dont l'un au moins n'est pas congru à $0 \pmod p$, on considère la forme de Klein $\mathfrak{k}_{(r,w)}$ (cf. [5], §1) :

$$\mathfrak{k}_{(r,w)}(z_1,z_2) = \exp\left(-\frac{r\underline{\eta}_1+w\underline{\eta}_2}{p}\,\frac{rz_1+wz_2}{2p}\right)\underline{\sigma}\left(\frac{rz_1+wz_2}{p};z_1,z_2\right),$$

où $\underline{\eta}_1, \underline{\eta}_2$ sont les quasi-périodes de la fonction zêta de Weierstrass associée au réseau engendré par z_1, z_2, et où $\underline{\sigma}$ est la fonction sigma de Weierstrass.

Nous considérerons des formes modulaires de la forme :

$$F(z) = \mathcal{K}\,.\prod_{(r,w)} \mathfrak{k}_{(r,w)}^{n(r,w)}(z)\,, \tag{2.3.1.1}$$

où $\mathcal{K} \in \mathbb{C}^\times$ est une constante, $n(r,w)$ un entier et où (r,w) parcourt un ensemble fini de couples d'entiers. La puissance 2p-ième de $\mathfrak{k}_{(r,w)}$ est une forme sur $\Gamma(p)$. Plus généralement, on a le résultat suivant :

Proposition 2.3.1.2 : Pour que la forme $F(z)$ définie par (2.3.1.1) soit une forme modulaire sur $\Gamma(p)$, il faut et il suffit que soient vérifiées les congruences suivantes :

$$\sum_{(r,w)} n(r,w)r^2 \equiv \sum_{(r,w)} n(r,w)rw \equiv \sum_{(r,w)} n(r,w).w^2 \equiv 0 \pmod p.$$

La démonstration est élémentaire à partir des propriétés K1, K2, K3 de [5], §1. (On rappelle que p est impair).

En particulier, $\mathfrak{k}^p_{(r,w)}$ est une forme sur $\Gamma(p)$.

Les formes $\mathfrak{k}_{(r,w)}$ possèdent un développement en produit eulérien (cf. [5], §1, K4).

2.3.2 Considérons l'ensemble $\amalg$:

$$\amalg = \{(r,w) \mid \left(\begin{smallmatrix} r\\ w\end{smallmatrix}\right) \in \Pi\}\,.$$

Le groupe H opère à droite sur $\amalg$. Soit $\varpi \in \amalg/H$ une orbite. On peut lui associer de façon canonique une forme F_ϖ en posant :

$$F_{\varpi}(z) = K_{\varpi} \cdot \prod_{(r,w)\in\varpi} \mathfrak{k}_{(r,w)}(z) \ ,$$

la constante K_{ϖ} étant choisie de telle sorte que le coefficient de la plus petite puissance de $q = e^{2\pi iz}$ dans le produit eulérien de $F_{\varpi}(z)$ soit 1. L'unicité résulte de ce que

$$\mathfrak{k}_{(-r,-w)} = -\mathfrak{k}_{(r,w)} \ ,$$

$$\mathfrak{k}_{(r+ap,w+bp)} = (-1)^{ab+a+b}\, e^{-2\pi i \frac{aw-br}{2p}}\, \mathfrak{k}_{(r,w)} \ ,$$

ce qui montre que le choix de représentants distincts dans $\mathbb{Z}^2$ de $(r,w) \in \amalg$ modifie le produit en le multipliant par une racine 2p-ième de l'unité.

Les formes F_{ϖ} vérifient de façon évidente, vu leur définition et les propriétés rappelées : Pour tout $\begin{pmatrix} a & b \\ c & d \end{pmatrix} \in \widetilde{H}$, on a

$$F_{\varpi}(az_1 + bz_2 \, , \, cz_1 + dz_2) = \chi_{\varpi}(\begin{pmatrix} a & b \\ c & d \end{pmatrix})\, F_{\varpi}(z_1, z_2) \ ,$$

où χ_{ϖ} est un caractère abélien de $\bar{H} = H/\{\pm 1\}$, à valeurs dans le groupe des racines 2p-ièmes de l'unité. (Par conséquent, χ_{ϖ} prend en fait ses valeurs dans l'intersection des groupes μ_{2p} et $\mu_{\frac{1}{2}|H|}$).

2.3.3 Considérons la pointe de X_H représentée par $\binom{u}{v} \in \prod$. Soit M une matrice de $SL_2(\mathbb{Z})$ telle que $M\binom{u}{v} = \binom{1}{0}$, et soit n l'entier défini en 2.2.2.

L'ordre de F_{ϖ} en la pointe considérée est alors donné par la formule suivante :

(2.3.3.1) $\quad$ ordre de F_{ϖ} en $\binom{u}{v} = \dfrac{n}{2p^2} \displaystyle\sum_{(r,w)\in\varpi.M^{-1}} r(r-p)$,

les entiers r étant choisis de façon à vérifier :

$$0 \leq r \leq \frac{p-1}{2} \ .$$

La formule (2.3.3.1) résulte facilement de l'expression du produit eulérien (cf. [5], § 1, K4).

2.3.4 Kubert et Lang montrent dans [5] le résultat suivant : Si le diviseur de F , fonction sur $X(p)$, est concentré aux pointes, il existe une puissance de F qui est un produit de formes de Klein $\mathfrak{k}_{(r,w)}$. Un résultat analogue vaut pour la courbe X_H et les formes F_{ϖ} , cf. [8]. Les calculs explicites que nous ferons au §4 constituent une vérification de ce résultat pour les courbes particulières étudiées.

3. Modèles sur $\mathbb{Q}$.

Le but de ce paragraphe est de définir des modèles sur $\mathbb{Q}$ de certaines des courbes X_H considérées aux paragraphes précédents. Pour cela, le point de vue naturel consiste à partir non pas d'un sous-groupe de $SL_2(\mathbb{F}_p)$, mais d'un sous-groupe de $GL_2(\mathbb{F}_p)$.

3.1 Considérons le corps $\mathbb{Q}(j,f_{(a,b)})$, où j désigne la fonction invariant modulaire, et où les $f_{(a,b)}$ sont les "Teilwerte" de niveau p (cf. [13], 6.6). Ce corps contient comme sous-corps algébriquement fermé le corps cyclotomique $\mathbb{Q}(\zeta_p)$, où $\zeta_p = e^{\frac{2\pi i}{p}}$.

Définition 3.1.1 : Le modèle canonique de $X(p)$ sur $\mathbb{Q}(\zeta_p)$ est la courbe propre et lisse sur $\mathbb{Q}(\zeta_p)$ dont le corps des fonctions est $\mathbb{Q}(j,f_{(a,b)})$.

Nous réservons dorénavant la notation $X(p)$ pour désigner le modèle canonique ainsi défini.

L'application qui, à $z \in \mathfrak{h}$ associe les valeurs en z de j et des $f_{(a,b)}$ définit une bijection canonique de $\mathfrak{h}^*/\Gamma(p)$ sur les points du $\mathbb{Q}(\zeta_p)$-schéma $X(p)$ à valeurs dans $\mathbb{C}$.

3.2 Soit $Y(p)$ l'ouvert de $X(p)$ formé du complémentaire des pointes. Du point de vue des schémas de modules, (cf. l'exposé de Deligne et Rapoport à Anvers (Lecture Notes n° 349)), le schéma $Y(p)$ représente le foncteur qui, à tout $\mathbb{Q}$-schéma S , associe l'ensemble des classes d'isomorphie des courbes elliptiques E/S munies d'un isomorphisme du noyau de la multiplication par p avec le schéma $(\mathbb{Z}/p\mathbb{Z})^2$. Le morphisme canonique de $Y(p)$ sur $\mathbb{Q}(\zeta_p)$, qui fait de $Y(p)$ un $\mathbb{Q}(\zeta_p)$-schéma, est défini par la forme de Weil.

Comme indiqué dans loc. cit., introd., il y a lieu de distinguer entre le $\mathbb{Q}$-schéma $Y(p)$ et le $\mathbb{Q}(\zeta_p)$-schéma $Y(p)$. En particulier, la fibre géométrique $Y(p)\otimes_{\mathbb{Q}}\bar{\mathbb{Q}}$ du premier est la somme disjointe de $(p-1)$ exemplaires de la fibre géométrique $Y(p)\otimes_{\mathbb{Q}(\zeta_p)}\bar{\mathbb{Q}}$, qui, elle, est lisse et connexe.

Les points du $\mathbb{Q}$-schéma $Y(p)$ à valeurs dans $\mathbb{C}$ paramétrisent les classes d'isomorphisme des triplets $(E,\binom{P}{Q})$ formés d'une courbe elliptique E et d'une base $\{P,Q\}$ du groupe des points de E annulés par p. Les points à valeurs dans $\mathbb{C}$ du $\mathbb{Q}(\zeta_p)$-schéma $Y(p)$ correspondent aux triplets vérifiant la condition supplémentaire :

$\langle P.Q\rangle = \zeta_p$, où $\langle\cdot\rangle$ désigne la forme bilinéaire de Weil.

L'application canonique de 3.1 s'interprète comme celle qui associe à $z\in\mathfrak{h}$ le triplet formé de la courbe elliptique définie par le réseau $\mathbb{Z}z+\mathbb{Z}$ et de la base $\binom{z/p}{1/p}$ des points annulés par p.

3.3 Le groupe $GL_2(\mathbb{F}_p)/\{\pm 1\}$ opère de façon naturelle sur le $\mathbb{Q}$-schéma $Y(p)$. En termes de triplets, la matrice $M\in GL_2(\mathbb{F}_p)$ associe au triplet :

$$(E,\binom{P}{Q}) \text{ le triplet } (E,M\binom{P}{Q}).$$

Cette action ne respecte pas la structure sur $\mathbb{Q}(\zeta_p)$. En termes de fonctions, l'action précédente correspond à l'action suivante sur $\mathbb{Q}(j,f_{(a,b)})$:

$$f_{(a,b)}|M = f_{(a,b)M}\ ,$$

et

$$\zeta_p|M = \zeta_p^{\det M}\ ,\ \text{cf. [13], th. 6.6 .}$$

3.4 Soit σ un automorphisme de $\mathbb{C}$. Soit $(E,\binom{P}{Q})$ un triplet représentant un point du $\mathbb{Q}$-schéma $Y(p)$ à valeurs dans $\mathbb{C}$. Choisissons une équation de E :

$$y^2 = x^3 - B_2x - B_3\ ,\quad B_2,B_3\in\mathbb{C}\ ;$$

le transformé de $(E,\binom{P}{Q})$ par σ est le triplet

$$(E^{\sigma},\binom{P^{\sigma}}{Q^{\sigma}}) ,$$

où E^{σ} désigne la courbe :

$$y^2 = x^3 - B_2^{\sigma}x - B_3^{\sigma} ,$$

et P^{σ} , Q^{σ} les points obtenus en faisant agir σ sur les coordonnées de P,Q .

En particulier, si σ fixe $\mathbb{Q}(\zeta_p)$, et si $\langle P.Q\rangle = \zeta_p$, on a $\langle P^{\sigma}.Q^{\sigma}\rangle = \zeta_p$, et ce qui précède décrit l'action galoisienne sur les points du $\mathbb{Q}(\zeta_p)$-schéma $Y(p)$ à valeurs dans $\mathbb{C}$.

Les pointes du modèle canonique $X(p)$ sont rationnelles sur $\mathbb{Q}(\zeta_p)$.

3.5 Soit G un sous-groupe de $GL_2(\mathbb{F}_p)$ contenant -1 .

Soit H l'intersection de G avec $SL_2(\mathbb{F}_p)$.

Notons $\mathbb{Q}(j,f_{(a,b)})^G$ le sous-corps de $\mathbb{Q}(j,f_{(a,b)})$ fixé par G , et soit $\mathbb{Q}_G = \mathbb{Q}(j,f_{(a,b)})^G \cap \mathbb{Q}(\zeta_p)$. Le corps $\mathbb{Q}_G$ est algébriquement fermé dans $\mathbb{Q}(j,f_{(a,b)})^G$.

<u>Définition</u> 3.5.1 : <u>On appelle modèle canonique associé à</u> G , <u>et on note</u> $X_G(p)$, <u>la courbe propre et lisse sur</u> $\mathbb{Q}_G$ <u>dont le corps des fonctions est</u> $\mathbb{Q}(j,f_{(a,b)})^G$.

Les points du $\mathbb{Q}_G$-schéma $X_G(p)$ à valeurs dans $\mathbb{C}$ qui ne sont pas des pointes paramétrisent les classes d'isomorphie d'orbites $(E,G\binom{P}{Q})$ qui contiennent un triplet $(E,\binom{P}{Q})$ vérifiant la condition $\langle P.Q\rangle = \zeta_p$.

De façon analogue à (3.1), on a une bijection canonique de $\mathfrak{h}^*/\widetilde{H}$ sur les pointes du $\mathbb{Q}_G$-schéma $X_G(p)$ à valeurs dans $\mathbb{C}$, où $\widetilde{H}$ (cf. §1) désigne l'image inverse de H dans $SL_2(\mathbb{Z})$.

3.6 Considérons le cas particulier où $\det(G) = \mathbb{F}_p^{\times}$, c'est-à-dire où la suite

$$1 \to H \to G \xrightarrow{\det} \mathbb{F}_p^{\times} \to 1$$

est exacte. Dans ce cas, $\mathbb{Q}_G = \mathbb{Q}$, et le modèle $X_G(p)$ est défini sur $\mathbb{Q}$. Les points de $X_G(p)$, autres que les pointes, paramétrisent les classes d'isomorphie d'orbites $(E,G(\substack{P\\Q}))$.

Soit σ un automorphisme de $\mathbb{C}$. Le transformé par σ de $(E,G(\substack{P\\Q}))$ est $(E^\sigma,G(\substack{P^\sigma\\Q^\sigma}))$. Si on a choisi dans $(E,(\substack{P\\Q}))$ un triplet vérifiant $\langle P.Q\rangle = \zeta_p$, on peut faire de même dans $(E^\sigma,G(\substack{P^\sigma\\Q^\sigma}))$. Plus précisément, soit m un générateur de $\mathbb{F}_p^\times$, et supposons que σ induise l'automorphisme $[m^a]$ (notation de I.2.4) sur $\mathbb{Q}(\zeta_p)$. Soit enfin $M \in G$ une matrice de déterminant m . Alors le triplet :

(3.6.1) $$(E^\sigma;M^{-a}(\substack{P^\sigma\\Q^\sigma}))$$

est dans l'orbite $(E^\sigma,G(\substack{P^\sigma\\Q^\sigma}))$, et vérifie la condition voulue.

Considérons enfin une pointe de $X_G(p)$, représentée par un élément $(\substack{u\\v}) \in \Pi$ (cf. 2.2.1). La pointe transformée par σ est représentée par :

(3.6.2) $$(\substack{u\\v})^{[m^a]} = M^{-a}.(\substack{u\\v}).$$

3.7 <u>Exemples</u>.

3.7.1 Supposons que G soit un groupe de Borel, resp. un groupe de Cartan (déployé ou non), resp. le normalisateur d'un groupe de Cartan de $GL_2(\mathbb{F}_p)$. Alors H est un groupe de même type de $SL_2(\mathbb{F}_p)$. On a $\det G = \mathbb{F}_p^\times$, donc $\mathbb{Q}_G = \mathbb{Q}$, et les modèles canoniques $X_G(p)$ correspondants sont des courbes sur $\mathbb{Q}$.

3.7.2 Soit G l'image réciproque dans $GL_2(\mathbb{F}_p)$ d'un sous-groupe de $PGL_2(\mathbb{F}_p)$ isomorphe au groupe symétrique $\mathfrak{S}_4$. Si $p \equiv 1 \pmod 4$, le sous-groupe de $PGL_2(\mathbb{F}_p)$ considéré est contenu dans $PSL_2(\mathbb{F}_p)$. On a alors $\mathbb{Q}_G = \mathbb{Q}(\sqrt{p})$. Si $p \equiv -1 \pmod 4$, $\det G = \mathbb{F}_p^\times$, et la courbe $X_G(p)$ associée à G est définie sur $\mathbb{Q}$.

3.7.3 Soit G le groupe formé des homothéties dans $GL_2(\mathbb{F}_p)$. La courbe correspondante est définie sur l'extension quadratique K_p de

$\mathbb{Q}$, où $K_p = \mathbb{Q}(\sqrt{p})$ si $p \equiv 1 \pmod 4$, $K_p = \mathbb{Q}(\sqrt{-p})$ sinon. Nous noterons cette courbe $X(p)_{K_p}$. C'est un K_p-modèle de $X(p)$ (cf. [13], 6.8). L'action de $GL_2(\mathbb{F}_p)$ sur le $\mathbb{Q}$-schéma $X(p)$ donne lieu à une action de $PSL_2(\mathbb{F}_p)$ sur le K_p-schéma $X(p)_{K_p}$, d'où une injection

(3.7.3.1) $$PSL_2(\mathbb{F}_p) \to \mathrm{Aut}_{K_p} X(p)_{K_p} .$$

3.7.4 Soit G le sous-groupe de $GL_2(\mathbb{F}_p)$ engendré par les matrices de la forme $\begin{pmatrix} * & 1 \\ 0 & 1 \end{pmatrix}$ et la matrice -1 . Le modèle associé à G est un $\mathbb{Q}$-modèle de la courbe $X_1(p)$. C'est le modèle considéré par Ogg dans [11].

3.7.5 Soit M une matrice de $GL_2(\mathbb{F}_p)$ satisfaisant aux deux conditions :

(i) $\det M$ engendre $\mathbb{F}_p^\times$;

(ii) $M^{p-1} = \pm 1$.

Soit G le plus petit groupe de $GL_2(\mathbb{F}_p)$ contenant M et -1. La courbe associée à G , que nous noterons $X_M(p)$, est un $\mathbb{Q}$-modèle de $X(p)$. Nous dirons que c'est le M-modèle de $X(p)$. Les formules 3.6.1 et 3.6.2 décrivent l'action d'un automorphisme σ de $\mathbb{C}$ sur un point de $X_M(p)$ à valeurs dans $\mathbb{C}$.

4. Les courbes $X_{\text{dép}}(11)$, $X_{\text{ndép}}(11)$, $X_{\mathfrak{S}_4}(11)$.

Nous sommes maintenant en mesure d'appliquer les techniques et résultats des paragraphes précédents au cas où $p = 11$.

4.1 Notations.

Soit G_o le groupe de Cartan déployé de $GL_2(\mathbb{F}_{11})$ formé des matrices diagonales.

Soit $G_{\text{dép}}$ le normalisateur de G_o dans $GL_2(\mathbb{F}_{11})$.

Soit $G_{\text{ndép}}$ le normalisateur du groupe de Cartan non déployé formé des matrices de $GL_2(\mathbb{F}_{11})$ de la forme

$$\begin{pmatrix} a & b \\ -b & a \end{pmatrix}, \quad a,b \in \mathbb{F}_{11}, \quad a^2+b^2 \neq 0 .$$

Soit $H_{\mathfrak{S}_4}$ le sous-groupe de $SL_2(\mathbb{F}_{11})$ engendré par les matrices :

$$h' = \begin{pmatrix} 0 & -3 \\ 4 & 0 \end{pmatrix}, \quad h'' = \begin{pmatrix} 5 & 5 \\ -4 & 5 \end{pmatrix}, \quad h''' = \begin{pmatrix} 2 & -4 \\ -2 & -1 \end{pmatrix} .$$

On a les relations :

$$h'^2 = -1 , \quad h''^3 = 1 , \quad h'''^3 = -1 , \quad h'h''h''' = -1 .$$

Il en résulte que $H_{\mathfrak{S}_4}$ contient -1, et que l'image $H_{\mathfrak{S}_4}/\{\pm 1\}$ de $H_{\mathfrak{S}_4}$ dans $PSL_2(\mathbb{F}_{11})$ est isomorphe au groupe alterné $\mathfrak{A}_4$.

La matrice $M = \begin{pmatrix} 3 & 0 \\ 0 & -3 \end{pmatrix}$ normalise $H_{\mathfrak{S}_4}$. Soit $G_{\mathfrak{S}_4}$ le sous-groupe de $GL_2(\mathbb{F}_{11})$ engendré par $H_{\mathfrak{S}_4}$ et M. Comme $-M^2$ engendre le groupe des homothéties, $G_{\mathfrak{S}_4}$ contient les homothéties. D'autre part, $\det(M) = 2$ engendre $\mathbb{F}_{11}^{\times}$; ceci entraîne que l'image de $G_{\mathfrak{S}_4}$ dans $PSL_2(\mathbb{F}_{11})$ contient $\mathfrak{A}_4$ comme sous-groupe normal d'indice 2, donc est isomorphe à $\mathfrak{S}_4$.

Soit H_o, resp. $H_{\text{dép}}$, resp. $H_{\text{ndép}}$, l'intersection de G_o, resp. $G_{\text{dép}}$, resp. $G_{\text{ndép}}$ avec $SL_2(\mathbb{F}_{11})$.

Nous noterons $X_{\text{dép}}(11)$, $X_{\text{ndép}}(11)$, $X_{\mathfrak{S}_4}(11)$ resp. les modèles canoniques associés aux groupes $G_{\text{dép}}$, $G_{\text{ndép}}$, $G_{\mathfrak{S}_4}$ (cf. 3.5.1). Ce sont des courbes sur $\mathbb{Q}$.

Remarquons que chacun des trois groupes $G_{\text{dép}}$, $G_{\text{ndép}}$, $G_{\mathfrak{S}_4}$ contient la matrice M. Les courbes associées apparaissent donc comme quotients sur $\mathbb{Q}$ du M-modèle $X_M(11)$ défini par M.

La formule 3.6.2 montre que l'automorphisme $[2] \in \mathrm{Gal}(\mathbb{Q}(\zeta_{11})/\mathbb{Q})$ agit de la façon suivante sur la pointe $\binom{u}{v} \in \Pi$ de $X_M(11)$:

$$\binom{u}{v}^{[2]} = \binom{4u}{-4v}. \tag{4.1.1}$$

Enfin, rappelons (cf. 2.1.2) que l'on note $\mathbb{E}_{11}$ le groupe $\mathbb{F}_{11}^{\times}/\{\pm 1\}$.

Passons à l'étude de chacune des trois courbes $X_{\text{dép}}(11)$, $X_{\text{ndép}}(11)$, $X_{\mathfrak{S}_4}(11)$.

4.2 <u>La courbe</u> $X_{\text{dép}}(11)$.

4.2.1 Les formules de 2.1 montrent que le genre de $X_{\text{dép}}(11)$ est 2. On est dans le cas où l'ordre du groupe H est premier à p. Par conséquent, $\bar{H}$ opère librement sur $\top\!\!\top$. On a donc $\frac{p^2-1}{|H|} = 6$ pointes. Ces pointes sont la pointe $P_{\text{dép},\infty}$, correspondant à l'orbite de $\binom{1}{0}$, et les pointes $P_{\text{dép},i}$ correspondant à l'orbite de $\binom{i}{1}$, pour $i \in E_{11}$.

La pointe $P_{\text{dép},\infty}$ est rationnelle sur $\mathbb{Q}$; d'autre part, on a, d'après 4.1.1 : $\qquad P^{[2]}_{\text{dép},i} = P_{\text{dép},6i}$, $i \in E_{11}$.

De même, à chacune des six orbites de l'action de $H_{\text{dép}}$ sur $\perp\!\!\perp$, qui sont celles de $(1,0)$ et de $(i,0)$, $i \in E_{11}$, correspond d'après 2.3.2 une forme modulaire, notée $F_{\text{dép},\infty}$ et $F_{\text{dép},i}$, $i \in E_{11}$, respectivement.

4.2.2 Nous n'utiliserons pas les résultats de [8] concernant l'ordre des pointes de $X_{\text{dép}}(p)$. Ces résultats entraîneraient ici que l'ordre des pointes divise 25.11. Nous allons au contraire procéder explicitement. Nous ferons de même pour $X_{\text{ndép}}(11)$ et $X_{\mathfrak{S}_4}(11)$.

Considérons d'abord l'ordre de $F_{\text{dép},\infty}$ aux différentes pointes. On a (cf. 2.3.3.1) :

$$\text{ordre de } F_{\text{dép},\infty} \text{ en } P_{\text{dép},\infty} = -5 \;;$$

$$\text{ordre de } F_{\text{dép},\infty} \text{ en } P_{\text{dép},i} = -10 \,,$$

$$\text{ordre de } F_{\text{dép},i} \text{ en } P_{\text{dép},\infty} = -10 \,,\ i \in E_{11} \,.$$

Considérons ensuite l'ordre de $F_{\text{dép},j}$ en la pointe $P_{\text{dép},i}$, $i,j \in E_{11}$. On a :

$$\text{ordre de } F_{\text{dép},j} \text{ en } P_{\text{dép},i} = \text{ordre de } F_{\text{dép},1} \text{ en } P_{\text{dép},ij} \,,\ i,j \in E_{11} \,,$$

et la première ligne de la matrice

$$(\text{ordre de } F_{\text{dép},j} \text{ en } P_{\text{dép},i})_{i,j\in \mathbb{E}_{11}}$$

est

$$(-9\,,\,-7\,,\,-9\,,\,-10\,,\,-10).$$

4.2.3 Considérons la forme :

$$F^{55}_{\text{dép},\infty}\, F^{-16}_{\text{dép},1}\, F^{-86}_{\text{dép},2}\, F^{-1}_{\text{dép},3}\, F^{19}_{\text{dép},4}\, F^{29}_{\text{dép},5}\ ;$$

cette forme définit sur $X_{\text{dép}}(11)$ une fonction dont le diviseur est

$$25.11(P_{\text{dép},\infty} - P_{\text{dép},1}).$$

Plongeons la courbe $X_{\text{dép}}(11)$ dans sa jacobienne $J_{\text{dép}}(11)$ en envoyant la pointe $P_{\text{dép},\infty}$ (qui est rationnelle sur $\mathbb{Q}$) sur l'origine.

4.2.4 Les propriétés de symétrie de la matrice donnant les ordres aux pointes entraînent :

La forme :

$$F^{55}_{\text{dép},\infty}\, F^{-16}_{\text{dép},\frac{1}{i}}\, F^{-86}_{\text{dép},\frac{2}{i}}\, F^{-1}_{\text{dép},\frac{3}{i}}\, F^{19}_{\text{dép},\frac{4}{i}}\, F^{29}_{\text{dép},\frac{5}{i}}\ ,\ i\in \mathbb{E}_{11}\ ,$$

définit sur $X_{\text{dép}}(11)$ une fonction dont le diviseur est :

$$25.11(P_{\text{dép},\infty} - P_{\text{dép},i})\ ,\ i\in \mathbb{E}_{11}\ .$$

Les pointes $P_{\text{dép},i}$, $i \in \mathbb{E}_{11}$ sont donc dans la jacobienne des points d'ordre divisant 25.11. Du reste, l'examen des développements en produit eulérien montre que cet ordre est exactement 25.11 (on utilise le fait que les coefficients de la série de Fourier ont des dénominateurs bornés ; cf. [7], dém. du lemme 3.2.15).

4.2.5 Considérons la forme :

$$F^{45}_{\text{dép},\infty}\, F^{-9}_{\text{dép},1}\, F^{-64}_{\text{dép},2}\, F^{1}_{\text{dép},3}\, F^{6}_{\text{dép},4}\, F^{21}_{\text{dép},5}\ .$$

Cette forme définit sur $X_{\text{dép}}(11)$ une fonction dont le diviseur est

$$25(9P_{\text{dép},\infty} - 8P_{\text{dép},1} - P_{\text{dép},5}).$$

D'autre part, la forme

Lig-56

$$F_{dép,\infty}^{11}\ F_{dép,1}^{-5}\ F_{dép,2}^{-20}\ F_{dép,3}^{-1}\ F_{dép,4}^{8}\ F_{dép,5}^{7}$$

définit une fonction de diviseur :

$$11(5P_{dép,\infty} - 6P_{dép,1} + P_{dép,5}).$$

4.2.6 De façon analogue à 4.2.4, on déduit l'existence sur $X_{dép}(11)$ de fonctions dont le diviseur est :

$$25(9P_{dép,\infty} - 8P_{dép,i} - P_{dép,5i}), \ i \in \mathbb{E}_{11} ;$$

$$11(5P_{dép,\infty} - 6P_{dép,i} + P_{dép,5i}), \ i \in \mathbb{E}_{11} .$$

4.2.7 Nous allons déduire de ce qui précède :

Proposition 4.2.7.1 : Le groupe cuspidal de $J_{dép}(11)$ est isomorphe, comme module galoisien, à la somme directe

$$C_2 \oplus C_6 ,$$

les modules galoisiens C_2, C_6 étant ceux considérés dans la prop. 2.6.1 du chapitre I.

En effet, soit $C_{dép}$ le groupe cuspidal de $J_{dép}(11)$, engendré par les pointes $P_{dép,i}$, $i \in \mathbb{E}_{11}$.

Considérons l'isomorphisme :

$$C_{dép} \to 11\ C_{dép} \times 25\ C_{dép} \quad \text{défini par :}$$

$$P \mapsto (11P , 25P),$$

et soit (Q_i', Q_i''), $i \in \mathbb{E}_{11}$, l'image de $P_{dép,i}$.

D'après 4.2.4, on a :

$$25Q_i' = 0 , \ 11Q_i'' = 0 , \ i \in \mathbb{E}_{11};$$

d'après 4.2.5 :

$$Q_{5i}' = 6Q_i' ,$$

$$Q_{5i}'' = -8\ Q_i'' , \qquad i \in \mathbb{E}_{11}.$$

Par conséquent, les groupes $11C_{dép}$, $25C_{dép}$ sont cycliques d'ordre 25 et 11 respectivement. D'autre part, si l'on se reporte à la définition de ε , cf. I, 2.4.1, les identités :

$$Q'_{5i} = 6Q'_i$$

s'écrivent :

$$Q'_i = \chi_{11}(i^4).\varepsilon(i^4)Q'_1 \ , \ i \in \mathbb{E}_{11}$$

Comme l'action du groupe de Galois est donnée par :

$$P^{[2]}_{\text{dép},i} = P_{\text{dép},6i} \ , \ \text{cf. } 4.2.1 \ ,$$

on a :

$$Q'^{[2]}_i = Q'_{6i} = \chi_{11}(9i^4)\varepsilon(9i^4)Q'_1 = \chi(2)\varepsilon(2)Q'_i \ ,$$

ce qui est précisément l'action sur le groupe $C_2 \subseteq E_2$, cf. I, prop. 2.6.1.

De même, les identités :

$$Q''_{5i} = -8 \ Q''_i$$

s'écrivent :

$$Q''_i = i^2.Q''_1 \ , \ i \in \mathbb{E}_n \ .$$

Comme

$$Q''^{[2]}_i = Q''_{6i} \ ,$$

on a :

$$Q''^{[2]}_i = 3i^2Q''_1 = 2^8.Q''_i \ ,$$

ce qui est l'action galoisienne sur C_6 , cf. I, prop. 2.6.1.

4.2.8 Nous allons démontrer le résultat suivant :

<u>Proposition</u> 4.2.8.1 : <u>Il existe une isogénie définie sur</u> $\mathbb{Q}$, <u>de noyau contenu dans le groupe des points de</u> $J_{\text{dép}}(11)$ <u>annulés par la multiplication par</u> 24 :

$$J_{\text{dép}}(11) \to E_2 \times E_6 \ ;$$

<u>la restriction de cette isogénie au groupe cuspidal est un isomorphisme</u> :

$$C_{dép} \xrightarrow{\sim} C_2 \times C_6 .$$

<u>Corollaire</u>. <u>La courbe</u> $X_{dép}(11)$ <u>n'a qu'un nombre fini de points rationnels sur</u> $\mathbb{Q}$.

En effet, d'après les tables de [2], on a : $E_2(\mathbb{Q}) \simeq \mathbb{Z}/5\mathbb{Z}$ et $E_6(\mathbb{Q}) \simeq \mathbb{Z}$. Considérons l'image $\bar{X}$ de $X_{dép}(11)$ dans le produit $E_2 \times E_6$. Il suffit de démontrer que $\bar{X}(\mathbb{Q})$ est fini. Considérons la projection $\bar{X} \to E_2$. Si $\bar{X}(\mathbb{Q})$ est infini, il existe une infinité de points rationnels dans l'une des cinq fibres au-dessus des cinq points rationnels de E_2 . Comme $\bar{X}$ engendre $E_2 \times E_6$, la projection $\bar{X} \to E_2$ est de degré fini, d'où une contradiction.

4.2.9 Remarquons tout d'abord que la courbe associée au groupe H_o , groupe diagonal de $SL_2(\mathbb{F}_{11})$, d'indice 2 dans $H_{dép}$, n'est autre que la courbe $X_o(121)$. En effet, l'application

$$z \mapsto 11z$$

du demi-plan de Poincaré complété $\mathfrak{H}^*$ dans lui-même induit un isomorphisme des surfaces de Riemann associées :

$$\psi_{\mathbb{C}} : \mathfrak{H}^*/\Gamma_o(121) \to \mathfrak{H}^*/\tilde{H}_o .$$

L'image inverse par $\psi_{\mathbb{C}}$ de la différentielle associée à une forme $f(z)$ est la différentielle associée à $11.f(11z)$.

Cet isomorphisme $\psi_{\mathbb{C}}$ provient d'un isomorphisme défini sur $\mathbb{Q}$. Plus précisément, un point de $X_o(121)$ autre qu'une pointe est représenté par un couple :

$$(E;C),$$

où E est une courbe elliptique, et C un sous-groupe de E cyclique d'ordre 121.

Soit alors ${}_{11}E$ le sous-groupe de E formé des points de E annulés par 11. Au couple (E,C), associons le triplet :

$$\psi(E,C) = (E/11C ; {}_{11}E/11C , C/11C).$$

Ce triplet est formé d'une courbe elliptique et de deux groupes cycliques d'ordre 11, dont la somme est le groupe des points annulés par 11. Un tel triplet représente un point de la courbe X_{H_o} associée à H_o. On obtient ainsi une application de $X(121)$ dans X_{H_o}, dont on vérifie qu'elle est définie sur $\mathbb{Q}$, et qu'elle coïncide sur $\mathbb{C}$ avec l'isomorphisme $\Psi_{\mathbb{C}}$.

De plus, la structure de X_{H_o} sur $\mathbb{Q}$ induite par le choix du modèle naturel de $X(11)$ coïncide avec la structure canonique de $X_o(121)$, lorsqu'on identifie X_{H_o} et $X_o(121)$ au moyen de Ψ.

4.2.10 La courbe $X_o(121)$ possède une involution, l'involution d'Atkin-Lehner W_{11} (cf. [1]), qui est un automorphisme de $X_o(121)$ défini sur $\mathbb{Q}$.

En termes de couples (E,C), cette involution correspond à l'application

$$(E;C) \mapsto (E/C, {}_{121}E/C).$$

En termes de surfaces de Riemann, W_{11} est définie par l'application:

$$z \mapsto -\frac{1}{121z} \quad \text{de } \mathfrak{h}^* \text{ dans lui-même.}$$

4.2.11 La courbe X_{H_o} admet elle aussi une involution naturelle V_{11}, définie sur $\mathbb{Q}$, qui correspond, en termes de triplets $(E;C',C'')$, formés d'une courbe elliptique E et de deux sous-groupes de E d'ordre 11, à l'échange de C' et C''.

En termes de surfaces de Riemann, V_{11} est définie par l'application:

$$z \mapsto -\frac{1}{z}$$

de $\mathfrak{h}^*$ dans lui-même.

4.2.12 Une vérification élémentaire, sur l'une ou l'autre des définitions de Ψ, montre que le diagramme

$$\begin{array}{ccc} X_o(121) & \xrightarrow{\psi} & X_{H_o} \\ W_{11} \downarrow & & \downarrow V_{11} \\ X_o(121) & \xrightarrow{\psi} & X_{H_o} \end{array}$$

est commutatif.

4.2.13 La courbe $X_{dép}(11)$ est le quotient de X_{H_o} par l'involution V_{11}. Elle s'identifie donc, grâce au morphisme ψ, au quotient de $X_o(121)$ par l'involution d'Atkin-Lehner W_{11}.

<u>Lemme</u> 4.2.13.1 : <u>Le sous-espace de</u> $\langle\Gamma_o(121),2\rangle_o$ <u>invariant par l'involution d'Atkin-Lehner</u> W_{11} <u>est l'espace engendré par les formes</u> f_2 <u>et</u> f_6 (cf. I, prop. 1.4.1).

Il suffit pour vérifier ce résultat d'utiliser les tables de [2]. On y trouve le fait que :

$$f_i|W_{11} = -f_i \text{ , pour } i = 3,4,5 \text{ , et } f_6|W_{11} = f_6 .$$

D'autre part, se reportant aux définitions de f_1 et f_2, et utilisant le fait que $f_1|\begin{pmatrix} 0 & -1 \\ 11 & 0\end{pmatrix} = -f_1$ (cf. [2] ou [7], 2.5.4), on obtient le résultat annoncé.

On déduit du lemme :

<u>Proposition</u> 4.2.13.2 : <u>Le quotient de</u> $J_o(121)$ <u>par l'involution d'Atkin-Lehner</u> W_{11} <u>est isogène sur</u> $\mathbb{Q}$ <u>au produit</u> $E_2 \times E_6$.

4.2.14 Démontrons maintenant la prop. 4.2.8.1. Tout d'abord :

<u>Lemme</u> 4.2.14.1 : <u>Soit</u> $\mathcal{G}$ <u>un groupe fini opérant sur une courbe algébrique propre et lisse</u> X. <u>Si tout élément de</u> $\mathcal{G}$ <u>possède au moins un point fixe, l'application canonique de la jacobienne de</u> $X/\mathcal{G}$ <u>dans celle de</u> X <u>est injective</u>.

La démonstration ne présente pas de difficultés.

Le lemme précédent s'applique à la courbe X_{H_o} et au groupe engendré par V_{11}. Par conséquent, la jacobienne $J_{dép}(11)$ s'injecte dans $J_o(121)$.

Plaçons-nous sur le corps des complexes, et considérons le diagramme commutatif :

$$\begin{array}{ccc} J_o(121) & \to & \mathbb{C}^6/L \\ \uparrow & & \downarrow \varphi_2 \times \varphi_6 \\ J_{dép}(11) & \to & E_2 \times E_6 \ . \end{array}$$

Le morphisme du bas est surjectif, et son noyau est contenu dans l'intersection de $J_{dép}(11)$ avec le noyau de l'isogénie canonique

$$\prod_{i=1}^{6} \varphi_i \ : \ J_o(121) \to \prod_{i=1}^{6} E_i \ ,$$

d'où le résultat, d'après le cor. de la prop. 4.9.7.1 du chap. I.

L'assertion concernant le groupe cuspidal est immédiate, compte tenu de la proposition 4.2.7.1.

4.3 <u>La courbe</u> $X_{ndép}(11)$.

4.3.1 Les formules de 2.1 montrent que la courbe $X_{ndép}(11)$ est de genre 1.

Les pointes de $X_{ndép}(11)$, notées $P_{ndép,i}$, $i \in \mathbb{E}_{11}$, correspondent aux cinq orbites de $\binom{u}{v}$, $u^2+v^2 = i$, $i \in \mathbb{E}_{11}$.

L'action du groupe de Galois est donnée par :

$$P^{[2]}_{ndép,i} = P_{ndép,5i} \ , \quad i \in \mathbb{E}_{11} \ , \quad \text{cf. } 4.1.1 \ .$$

Les cinq orbites de l'action de $H_{ndép}$ sur $\coprod$ sont celles de (r,w), $r^2+w^2 = j$, $j \in \mathbb{E}_{11}$. A chacune est associée une forme modulaire notée $F_{ndép,j}$.

4.3.2 Considérons l'ordre des formes $F_{ndép,j}$ aux pointes $P_{ndép,i}$. On a (cf. 2.3.3.1) :

ordre de $F_{ndép,j}$ en $P_{ndép,i}$ = ordre de $F_{ndép,1}$ en $P_{ndép,ij}$, $i,j \in \mathbb{E}_{11}$,

et la première ligne de la matrice :

$$(\text{ordre de } F_{ndép,j} \text{ en } P_{ndép,i})_{i,j \in \mathbb{E}_{11}}$$

est

(-11 , -10 , -13 , -11 , -10).

4.3.3 Considérons la forme :

$$F^{-6}_{ndép,1}\ F^{9}_{ndép,2}\ F^{-21}_{ndép,3}\ F^{14}_{ndép,4}\ F^{4}_{ndép,5}\ ;$$

elle définit sur $X_{ndép}(11)$ une fonction dont le diviseur est :

$$5.11(P_{ndép,1} - P_{ndép,2}).$$

Les propriétés de symétrie de la matrice donnant les ordres aux pointes entraînent :

La forme

$$F^{-6}_{ndép,\frac{1}{i}}\ F^{9}_{ndép,\frac{2}{i}}\ F^{-21}_{ndép,\frac{3}{i}}\ F^{14}_{ndép,\frac{4}{i}}\ F^{4}_{ndép,\frac{5}{i}}$$

définit sur $X_{ndép}(11)$ une fonction dont le diviseur est :

$$5.11(P_{ndép,i} - P_{ndép,2i}).$$

4.3.4 Considérons maintenant le morphisme ι de $X_{ndép}(11)$ dans sa jacobienne $J_{ndép}(11)$ défini de la façon suivante :

$$\iota : X_{ndép}(11) \to J_{ndép}(11)$$
$$P \mapsto 5P - \sum_{i=1}^{5} P_{ndép,i} = \iota(P).$$

Le morphisme ι est défini sur $\mathbb{Q}$, parce que le diviseur $\sum_{i=1}^{5} P_{ndép,i}$ est rationnel sur $\mathbb{Q}$.

4.3.5 Considérons la forme :

$$F^{2}_{ndép,2}\ F^{-4}_{ndép,3}\ F^{-1}_{ndép,4}\ F^{3}_{ndép,5}\ ;$$

elle définit sur $X_{ndép}(11)$ une fonction de diviseur :

$$13P_{ndép,1} - 2P_{ndép,2} - 2P_{ndép,3} - 2P_{ndép,4} - 7P_{ndép,5}$$

soit encore

$$= 3\iota(P_{ndép,1}) - \iota(P_{ndép,5}).$$

On en déduit par permutation des fonctions de diviseur

$$3\iota(P_{ndép,i}) - \iota(P_{ndép,5i}),\ i \in \mathbb{E}_{11} .$$

4.3.6 Considérons enfin les fonctions :

$$F^{2}_{\mathrm{ndép},\frac{1}{i}}\ F^{8}_{\mathrm{ndép},\frac{2}{i}}\ F^{-15}_{\mathrm{ndép},\frac{3}{i}}\ F^{-1}_{\mathrm{ndép},\frac{4}{i}}\ F^{6}_{\mathrm{ndép},\frac{5}{i}}\ ,\ i \in \mathbb{E}_{11}\ ;$$

elles définissent sur $X_{\mathrm{ndép}}$ des fonctions de diviseurs

$$11(4P_{\mathrm{ndép},i} - P_{\mathrm{ndép},2i} - P_{\mathrm{ndép},3i} - P_{\mathrm{ndép},4i} - P_{\mathrm{ndép},5i}),$$

$$= 11.\iota(P_{\mathrm{ndép},i}).$$

4.3.7 D'après 4.3.5, 4.3.6, le groupe engendré dans $J_{\mathrm{ndép}}(11)$ par les images des pointes est cyclique d'ordre 11. Plus précisément, on déduit de 4.3.5 :

$$\iota(P_{\mathrm{ndép},i}) = i^{2}.\iota(P_{\mathrm{ndép},1}).$$

L'action du groupe de Galois commute à ι , et on a :

$$\iota(P_{\mathrm{ndép},i})^{[2]} = \iota(P_{\mathrm{ndép},5i}) = 3.i^{2}\iota(P_{\mathrm{ndép},1}) = 2^{8}.\iota(P_{\mathrm{ndép},i}),$$

ce qui montre que l'image du groupe cuspidal par ι est isomorphe comme module galoisien au groupe $C_6 \cong \mu_{11}^{\otimes 8}$.

La prop. 5.6.1 du chap. I s'applique à $J_{\mathrm{ndép}}(11)$, et entraîne :

$J_{\mathrm{ndép}}(11)$ est isomorphe sur $\mathbb{Q}$ à la courbe E_6 .

4.3.8 On sait construire des points de $X_{\mathrm{ndép}}(11)$ rationnels sur $\mathbb{Q}$, par exemple au moyen de courbes d'invariant $j=0$ ou 1728. Par conséquent, la courbe $X_{\mathrm{ndép}}(11)$ est isomorphe sur $\mathbb{Q}$ à sa jacobienne, donc à E_6 :

<u>Proposition</u> 4.3.8.1 : <u>La courbe</u> $X_{\mathrm{ndép}}(11)$ <u>est isomorphe sur</u> $\mathbb{Q}$ <u>à la courbe</u> E_6 , <u>d'équation</u> :

$$y^{2}+y = x^{3}-x^{2}-7x+10 \quad \text{(cf. I, 2.6 et [2]).}$$

<u>Corollaire</u>. <u>Le groupe des points de</u> $X_{\mathrm{ndép}}(11)$ <u>rationnels sur</u> $\mathbb{Q}$ <u>est isomorphe à</u> $\mathbb{Z}$.

En effet, $E_6(\mathbb{Q}) \cong \mathbb{Z}$, d'après les tables de [2].

4.4 <u>La courbe</u> $X_{\mathfrak{G}_4}(11)$.

4.4.1 Les formules de 2.1 montrent que la courbe $X_{\mathfrak{G}_4}(11)$ est de genre 1.

Les pointes de $X_{\mathfrak{G}_4}(11)$, notées $P_{\mathfrak{G}_4,i}$, $i \in \mathbb{E}_{11}$, correspondent aux cinq orbites de $\binom{i}{0}$, $i \in \mathbb{E}_{11}$.

L'action du groupe de Galois est donnée par :

$$P^{[2]}_{\mathfrak{G}_4,i} = P_{\mathfrak{G}_4,4i} \ , \ i \in \mathbb{E}_{11} \ , \quad \text{cf. } 4.1.1 .$$

Les cinq orbites de l'action de $H_{\mathfrak{G}_4}$ sur $\amalg$ sont celles de $(j,0)$, $j \in \mathbb{E}_{11}$. A chacune est associée une forme modulaire notée $F_{\mathfrak{G}_4,j}$.

4.4.2 Considérons l'ordre des formes $F_{\mathfrak{G}_4,j}$ aux pointes $P_{\mathfrak{G}_4,i}$. On a (cf. 2.3.3.1) :

$$\text{ordre de } F_{\mathfrak{G}_4,j} \text{ en } P_{\mathfrak{G}_4,i} = \text{ordre de } F_{\mathfrak{G}_4,1} \text{ en } P_{\mathfrak{G}_4,ij} \ , \ i,j \in \mathbb{E}_{11} ,$$

et la première ligne de la matrice :

$$(\text{ordre de } F_{\mathfrak{G}_4,j} \text{ en } P_{\mathfrak{G}_4,i})_{i,j \in \mathbb{E}_{11}}$$

est $$(-11 \ , -9 \ , -12 \ , -12 \ , -11).$$

4.4.3 Considérons la forme :

$$F^{-17}_{\mathfrak{G}_4,1} \ F^{23}_{\mathfrak{G}_4,2} \ F^{-7}_{\mathfrak{G}_4,3} \ F^{-2}_{\mathfrak{G}_4,4} \ F^{3}_{\mathfrak{G}_4,5} \ ;$$

elle induit sur $X_{\mathfrak{G}_4}(11)$ une fonction de diviseur :

$$5.11(P_{\mathfrak{G}_4,1} - P_{\mathfrak{G}_4,2}).$$

Les propriétés de symétrie de la matrice donnant les ordres aux pointes entraînent :

La forme

$$F^{-17}_{\mathfrak{G}_4,\frac{1}{i}} \ F^{23}_{\mathfrak{G}_4,\frac{2}{i}} \ F^{-7}_{\mathfrak{G}_4,\frac{3}{i}} \ F^{-2}_{\mathfrak{G}_4,\frac{4}{i}} \ F^{3}_{\mathfrak{G}_4,\frac{5}{i}} \ , \ i \in \mathbb{E}_{11} ,$$

induit sur $X_{\mathfrak{G}_4}(11)$ une fonction de diviseur

$$5.11(P_{\mathfrak{G}_4,i} - P_{\mathfrak{G}_4,2i}).$$

4.4.4 Considérons le morphisme ι de $X_{\mathfrak{S}_4}(11)$ dans sa jacobienne $J_{\mathfrak{S}_4}(11)$, défini de la façon suivante :

$$\iota : X_{\mathfrak{S}_4}(11) \to J_{\mathfrak{S}_4}(11)$$
$$P \mapsto 5P - \sum_{i=1}^{5} P_{\mathfrak{S}_4,i} = \iota(P).$$

Ce morphisme est défini sur $\mathbb{Q}$ (cf. 4.3.4).

4.4.5 La forme :

$$F^{6}_{\mathfrak{S}_4,2}\, F^{-2}_{\mathfrak{S}_4,3}\, F^{-3}_{\mathfrak{S}_4,4}\, F^{-1}_{\mathfrak{S}_4,5}$$

définit sur $X_{\mathfrak{S}_4}(11)$ une fonction de diviseur

$$17P_{\mathfrak{S}_4,1} - 3P_{\mathfrak{S}_4,2} - 3P_{\mathfrak{S}_4,3} - 8P_{\mathfrak{S}_4,4} - 3P_{\mathfrak{S}_4,5}$$
$$= 4\iota(P_{\mathfrak{S}_4,1}) - \iota(P_{\mathfrak{S}_4,4}).$$

On en déduit par permutation des fonctions de diviseurs

$$4\iota(P_{\mathfrak{S}_4,i}) - \iota(P_{\mathfrak{S}_4,4i}), \quad i \in \mathbb{E}_{11}.$$

4.4.6 Enfin, les formes :

$$F^{-2}_{\mathfrak{S}_4,\frac{1}{i}}\, F^{15}_{\mathfrak{S}_4,\frac{2}{i}}\, F^{-6}_{\mathfrak{S}_4,\frac{3}{i}}\, F^{-8}_{\mathfrak{S}_4,\frac{4}{i}}\, F^{1}_{\mathfrak{S}_4,\frac{5}{i}}, \quad i \in \mathbb{E}_{11},$$

induisent sur $X_{\mathfrak{S}_4}(11)$ des fonctions de diviseurs :

$$11(4P_{\mathfrak{S}_4,i} - P_{\mathfrak{S}_4,2i} - P_{\mathfrak{S}_4,3i} - P_{\mathfrak{S}_4,4i} - P_{\mathfrak{S}_4,5i})$$
$$= 11.\iota(P_{\mathfrak{S}_4,i}).$$

4.4.7 Il résulte de 4.4.5 et 4.4.6 que le groupe engendré dans $J_{\mathfrak{S}_4}(11)$ par les images des pointes est cyclique d'ordre 11. Plus précisément, on déduit de 4.4.5 :

$$\iota(P_{\mathfrak{S}_4,i}) = i^6 . \iota(P_{\mathfrak{S}_4,1}), \quad i \in \mathbb{E}_{11}.$$

L'action du groupe de Galois commute à ι, et on a :

$$\iota(P_{\mathfrak{S}_4,i})^{[2]} = \iota(P_{\mathfrak{S}_4,4i}) = 4.i^6.\iota(P_{\mathfrak{S}_4,1}) = 2^2.\iota(P_{\mathfrak{S}_4,i}),$$

ce qui montre que l'image du groupe cuspidal par ι est isomorphe comme module galoisien au groupe $\mu_{11}^{\otimes 2}$.

La prop. 5.6.1 du chap. I s'applique à $J_{\mathfrak{C}_4}(11)$, et entraîne :

$J_{\mathfrak{C}_4}(11)$ est isomorphe sur $\mathbb{Q}$ à la courbe $E'_4 = E_5/C_5$.

4.4.8 Un point rationnel sur $\mathbb{Q}$ de la courbe $X_{\mathfrak{C}_4}(11)$ a été construit explicitement par J.-P. Serre. Par conséquent, la courbe $X_{\mathfrak{C}_4}(11)$ est isomorphe sur $\mathbb{Q}$ à sa jacobienne :

<u>Proposition</u> 4.4.8.1 : <u>La courbe</u> $X_{\mathfrak{C}_4}(11)$ <u>est isomorphe sur</u> $\mathbb{Q}$ <u>à la courbe</u> $E'_4 \cong E_5/C_5$, <u>d'équation</u> :

$$y^2+xy+y = x^3+x^2-305x+7888 \quad \text{(cf. I, 2.6 et [2])}.$$

<u>Corollaire</u>. <u>La courbe</u> $X_{\mathfrak{C}_4}(11)$ <u>possède un seul point rationnel sur</u> $\mathbb{Q}$.

En effet, les tables de [2] montrent que $E'_4(\mathbb{Q})$ contient un seul élément.

III. Représentations de Hecke

1. Exposé du problème.

1.1 Soit $\mathcal{V} = \langle\Gamma(p),2\rangle_o$ l'espace des formes paraboliques de poids 2 sur le groupe $\Gamma(p)$.

Le groupe $SL_2(\mathbb{Z})$ opère sur $\mathcal{V}$ de la façon habituelle :

Si $f \in \mathcal{V}$, et $M = \binom{a\ b}{c\ d} \in SL_2(\mathbb{Z})$,

$(f|M)(z) = (cz+d)^{-2}.f(\frac{az+b}{cz+d})$; le sous-groupe $\{\pm 1\}.\Gamma(p)$ opère trivialement, d'où une action du groupe quotient

$$\mathcal{G}_p = PSL_2(\mathbb{F}_p) \quad \text{sur} \quad \mathcal{V}.$$

Nous dirons que $\mathcal{V}$, munie de l'action de $\mathcal{G}_p$, est la représentation de Hecke de $\mathcal{G}_p$.

1.2 Hecke a considéré dans [3], puis dans [4], le problème suivant:

Problème. Déterminer la décomposition de la représentation $\mathcal{V}$ de $\mathcal{G}_p$ en somme de représentations irréductibles.

(En fait, le second article traite du cas plus général où l'on remplace $\Gamma(p)$ par un sous-groupe arbitraire de $SL_2(\mathbb{Z})$, normal et d'indice fini dans $SL_2(\mathbb{Z})$).

Le problème se subdivise en deux problèmes de natures distinctes, à savoir :

1) Soit $\mathfrak{D}$ une représentation irréductible de $\mathcal{G}_p$, et soit $\bar{\mathfrak{D}}$ la représentation conjuguée de $\mathfrak{D}$. Soient $r_{\mathfrak{D}}$, $r_{\bar{\mathfrak{D}}}$ les multiplicités respectives de $\mathfrak{D}$ et $\bar{\mathfrak{D}}$ dans $\mathcal{V}$. Déterminer $r_{\mathfrak{D}} + r_{\bar{\mathfrak{D}}}$.

Ce problème est de nature topologique.

2) Si $\mathfrak{D}$ et $\bar{\mathfrak{D}}$ ne sont pas isomorphes, déterminer $r_{\mathfrak{D}}$.

Ce deuxième problème fait intervenir des considérations de nature arithmétique.

1.3 Les deux articles mentionnés plus haut fournissent chacun une méthode pour aborder le problème 1.

1.3.1 <u>Premier problème</u>. Soit $\bar{H}$ un sous-groupe de $\mathcal{G}_p$. Soit $\mathcal{V}^H$ le sous-espace de $\mathcal{V}$ invariant sous H .

On a :

$$\dim \mathcal{V}^H = \text{genre de la courbe } X_H$$

$$= \sum_{\mathfrak{D}} r_{\mathfrak{D}} . \dim \mathfrak{D}^H ,$$

où $\mathfrak{D}^H$ désigne le sous-espace de $\mathfrak{D}$ invariant sous H .

On sait déterminer le genre de X_H (lorsque H est cyclique, ceci est fait dans [3] ; cf. également II, 2.1).

D'autre part :

$$\dim \mathfrak{D}^H = \frac{1}{|H|} \sum_{h \in H} \chi_{\mathfrak{D}}(h) ,$$

où $\chi_{\mathfrak{D}}$ désigne le caractère de la représentation irréductible $\mathfrak{D}$.

Si l'on porte cette expression dans celle obtenue pour $\dim \mathcal{V}^H$, et si l'on fait parcourir à $\bar{H}$ l'ensemble des sous-groupes cycliques de $\mathcal{G}_p$, on obtient un système linéaire d'équations aux inconnues $r_{\mathfrak{D}}$. On sait (cf. [12], II.13.1, cor. du th. 30') qu'une représentation dont le caractère est à valeurs dans $\mathbb{Q}$ est connue lorsqu'on connait les invariants des sous-groupes cycliques. La résolution du système donne donc, pour chaque représentation irréductible $\mathfrak{D}$ de $\mathcal{G}_p$, la valeur de $\sum_{\mathfrak{D}'} r_{\mathfrak{D}'}$, $\mathfrak{D}'$ parcourant les représentations conjuguées de $\mathfrak{D}$ sur $\mathbb{Q}$. A cause des propriétés particulières de $\mathcal{G}_p$ (cf. 1.4) et de la représentation $\mathcal{V}$ (cf. prop. 3.1.1), ceci équivaut à la résolution du problème 1.

Un inconvénient de cette méthode est que la détermination de $r_{\mathfrak{D}}$, pour $\mathfrak{D}$ donné, exige la connaissance des caractères $\chi_{\mathfrak{D}'}$, pour tous les $\mathfrak{D}'$.

1.3.2 <u>Deuxième méthode</u>. L'espace $\mathcal{V}$ s'identifie de façon canonique avec l'espace des différentielles de première espèce sur la surface de Riemann $X(p)$. L'idée de Hecke, mise en oeuvre dans [4], consiste à considérer, plutôt que les différentielles, les intégrales de première espèce.

On obtient ainsi une représentation affine du groupe $\mathcal{G}_p$ où, en quelque sorte, les $r_{\mathfrak{D}}$ copies d'une représentation $\mathfrak{D}$ intervenant avec la multiplicité $r_{\mathfrak{D}}$ sont séparées.

Le résultat principal est le suivant ([4], th. 8) : Désignons par S, resp. T les matrices $\begin{pmatrix} 0 & -1 \\ 1 & 0 \end{pmatrix}$, resp. $\begin{pmatrix} 1 & 1 \\ 0 & 1 \end{pmatrix}$.

<u>Proposition</u> 1.3.2.1 : <u>Soit</u> $\mathfrak{D}$ <u>une représentation irréductible de</u> $\mathcal{G}_p$, <u>distincte de la représentation unité</u>. <u>Soit</u> $\chi_{\mathfrak{D}}$ <u>son caractère</u>. <u>Alors</u> :

$$r_{\mathfrak{D}} + r_{\bar{\mathfrak{D}}} = \deg \mathfrak{D} - \frac{1}{p} \sum_{a \bmod p} \chi_{\mathfrak{D}}(T^a) - \frac{1}{2} \sum_{a \bmod 2} \chi_{\mathfrak{D}}(S^a) - \frac{1}{3} \sum_{a \bmod 3} \chi_{\mathfrak{D}}((ST)^a).$$

Le premier problème posé est ainsi entièrement résolu. De plus, la connaissance de $r_{\mathfrak{D}} + r_{\bar{\mathfrak{D}}}$ ne demande que celle du caractère $\chi_{\mathfrak{D}}$ de la représentation $\mathfrak{D}$ considérée.

1.4 <u>Représentations irréductibles de</u> $\mathcal{G}_p$.

Pour résoudre le second problème, on doit avoir plus d'informations sur les représentations irréductibles de $\mathcal{G}_p$. En particulier, on désire savoir lesquelles sont réelles.

1.4.1 On connaît, grâce à Frobenius, toutes les représentations irréductibles de $\mathcal{G}_p = PSL_2(\mathbb{F}_p)$, (cf. [3], §1). Elles sont au nombre de $\frac{p+5}{2}$.

Mise à part la représentation unité, ce sont les représentations données par le tableau suivant :

degré	p	$\frac{p+(-1)^{\frac{p-1}{2}}}{2} = k$	p+1	p-1
représentation	1 représ. irréductible $\mathfrak{S}_p$	2 représ. conjuguées $\mathfrak{S}_k$ $\bar{\mathfrak{S}}_k$	pour chaque diviseur t de $\frac{p-1}{2}$, $t>2$, $\frac{1}{2}\varphi(t)$ représ. $\mathfrak{S}^{(a)}_{p+1}(t)$	pour chaque diviseur t de $\frac{p+1}{2}$, $t>2$, $\frac{1}{2}\varphi(t)$ représ. $\mathfrak{S}^{(a)}_{p-1}(t)$
caractère	réel (rationnel)	réel ou non selon que $p\equiv 1$ ou 3 (mod 4) (à valeurs dans K_p)	réel (à valeurs dans $\mathbb{Q}(\zeta_t+\zeta_t^{-1})$)	réel (à valeurs dans $\mathbb{Q}(\zeta_t+\zeta_t^{-1})$)

1.4.1.2 1) Considérons la représentation naturellement associée à l'action de G_p sur la droite projective $\mathbb{P}^1(\mathbb{F}_p)$. Elle contient une sous-représentation irréductible de degré p qui est $\mathfrak{S}_p$. Cette dernière est donc réalisable sur $\mathbb{Q}$.

2) On peut démontrer ([4], Satz 14) que toute représentation irréductible de G_p est réalisable sur le corps engendré par son caractère.

3) Lorsque $p\equiv 3$ (mod 4), on sait construire explicitement au moyen de séries thêta, un sous-espace invariant de $\mathcal{V}$ qui est somme directe d'un certain nombre (égal au nombre de classes du corps $\mathbb{Q}(\sqrt{-p})$) d'exemplaires de la représentation $\mathfrak{S}_{\frac{p-1}{2}}$ (cf. [3], §4). Nous allons expliciter cette construction dans le cas où p = 11 (cf. 2.2).

2. Le cas p = 11 .

2.1 Lorsque p = 11 , la représentation $\mathcal{V}$ est de dimension 26. Utilisons la prop. 1.3.2.1. Les valeurs de $\chi_{\mathfrak{S}}$ sont données par le tableau de [3], §1. On trouve :

	rep. unité	$\mathcal{D}_{11}$	$\mathcal{D}_5$ ou $\bar{\mathcal{D}}_5$	$\mathcal{D}_{10}(3)$	$\mathcal{D}_{10}(6)$	$\mathcal{D}_{12}^{(1)}(5)$	$\mathcal{D}_{12}^{(2)}(5)$
$r_{\mathcal{D}} + r_{\bar{\mathcal{D}}}$	0	1	1	1	0	0	0

Il reste à résoudre le problème 2, c'est-à-dire ici à déterminer laquelle des représentations $\mathcal{D}_5$, $\bar{\mathcal{D}}_5$ intervient dans $\mathcal{V}$.

2.2 Considérons les séries thêta :

$$\theta(z,j,\sqrt{-11}) = \sum_{\substack{\alpha \in O_K \\ \alpha \equiv j (\mathrm{mod}\ \sqrt{-11})}} \alpha . e^{2\pi i z \frac{\alpha\bar{\alpha}}{11}} , \quad j = 1,\ldots,5 .$$

Ces séries engendrent une sous-représentation de dimension 5 de $\mathcal{V}$, isomorphe à $\mathcal{D}_5$. On en déduit que $r_{\mathcal{D}_5} = 1$, $r_{\bar{\mathcal{D}}_5} = 0$.

2.3 En définitive, la représentation $\mathcal{V}$ est isomorphe à la somme directe :

$$\mathcal{D}_{11} \oplus \mathcal{D}_5 \oplus \mathcal{D}_{10}(3).$$

Les représentations $\mathcal{D}_{11}$, $\mathcal{D}_{10}(3)$ sont réalisables sur $\mathbb{Q}$. La représentation $\mathcal{D}_5$ est réalisable sur $K = \mathbb{Q}(\sqrt{-11})$.

2.4 Considérons les sous-groupes $H_{\text{dép}}$, $H_{\text{ndép}}$, $H_{\mathfrak{S}_4}$ (cf. II,4.1). Utilisant à nouveau la connaissance explicite des caractères des représentations irréductibles, et la prop. 1.3.2.1, on calcule la dimension de l'espace des invariants de chacun de ces groupes dans les représentations irréductibles $\mathcal{D}_{11}$, $\mathcal{D}_5$, $\mathcal{D}_{10}(3)$. Le calcul ne présente pas de difficultés. On obtient le résultat suivant :

	$\mathcal{D}_{11}$	$\mathcal{D}_5$	$\mathcal{D}_{10}(3)$
$H_{\text{dép}}$	1	1	0
$H_{\text{ndép}}$	0	1	0
$H_{\mathfrak{S}_4}$	0	0	1

3. Interprétation géométrique.

3.1 Reprenons les notations de II, §§3 et 4. Rappelons (cf. II, ex. 3.7.3) que $X(p)_{K_p}$ note le modèle canonique associé au sous-groupe de $GL_2(\mathbb{F}_p)$ formé des homothéties, et que l'on a une injection canonique :

$$\mathcal{G}_p = PSL_2(\mathbb{F}_p) \rightarrow Aut_{K_p}(X(p)_{K_p}) .$$

On notera $J(p)_{K_p}$ la jacobienne de $X(p)_{K_p}$. On déduit du morphisme précédent un morphisme canonique :

$$\mathcal{G}_p \rightarrow Aut_{K_p}(J(p)_{K_p}) .$$

Soit D_{K_p} l'espace des 1-différentielles invariantes sur la variété abélienne $J(p)_{K_p}$. Le morphisme que l'on vient de définir induit un morphisme

$$\mathcal{G}_p \rightarrow Aut_{K_p}(D_{K_p}) ,$$

qui fait de D_{K_p} une représentation K_p-linéaire de $\mathcal{G}_p$. Cette représentation est une réalisation sur K_p de la représentation de Hecke $\mathcal{V}$. Par conséquent :

Proposition 3.1.1 : La représentation de Hecke $\mathcal{V}$ est réalisable sur le corps K_p.

3.2 On suppose maintenant que $p = 11$. On notera simplement M l'automorphisme de $J(11)_K$ associé à $M \in \mathcal{G}_{11}$.

3.2.1 Soit H un sous-groupe de $SL_2(\mathbb{F}_{11})$, et supposons que la courbe X_H associée soit de genre 1. Les invariants de H forment donc un espace de dimension 1 dans une représentation irréductible $\mathcal{S}_d$ de $\mathcal{G} = \mathcal{G}_{11}$.

Soit f une base de l'espace des invariants de H. La représentation $\mathcal{S}_d$ étant irréductible, on a :

$$f|\mathcal{G} = \mathcal{S}_d .$$

On peut donc choisir d éléments $M_1 = 1$, $M_2,\ldots,M_d$ de $\mathcal{G}$ tels que les formes :

$$f_{M_i} = f|M_i \quad , \; i = 1,\ldots,d \; ,$$

soient une base de $\mathcal{S}_d$.

Considérons alors les sous-groupes :

$$H_i = M_i^{-1} H M_i \; , \; i = 1,\ldots,d \; .$$

Le vecteur f_{M_i} est une base de l'espace des invariants sous le groupe H_i .

Soit X_H , resp. X_{H_i} la courbe quotient de $X(11)_K$ par le groupe d'automorphismes défini par H , resp. H_i . Remarquons que, si H est l'un des groupes considérés en 4.1, la courbe X_H s'obtient à partir du modèle canonique défini en 4.1 par extension des scalaires de $\mathbb{Q}$ à K . L'automorphisme de $X(11)_K$ défini par M_i définit un isomorphisme de X_H avec X_{H_i} :

$$X_{H_i} \xrightarrow{\sim} X_H \; .$$

3.2.2 Considérons le K-morphisme

$$X(11)_K \to X_{H_1} \times\ldots\times X_{H_d} \simeq X_H^d$$

produit des morphismes canoniques.

On en déduit un morphisme des jacobiennes :

$$J(11)_K \to J_H^d \; .$$

Ce morphisme est surjectif. En effet, l'application cotangente correspond à l'injection du sous-espace engendré par les f_{M_i} , c'est-à-dire de $\mathcal{S}_d$ dans l'espace $\mathcal{V}$ lui-même.

3.2.3 En définitive, on a montré :

<u>La décomposition de</u> $J(11)_K$ <u>en classes d'isogénies sur</u> K <u>contient</u> d <u>exemplaires de</u> J_H .

3.3 Ce qui précède s'applique dans les trois cas suivants, (cf. tableau de 2.4) :

1) Le groupe H est le groupe de Borel de $SL_2(\mathbb{F}_{11})$ formé des matrices triangulaires supérieures, et $\mathfrak{D} = \mathfrak{D}_{11}$.

2) Le groupe H est $H_{ndép}$, et $\mathfrak{D} = \mathfrak{D}_5$.

3) Le groupe H est $H_{\mathfrak{S}_4}$, et $\mathfrak{D} = \mathfrak{D}_{10}(3)$.

On obtient donc :

La décomposition de $J(11)_K$ en classes d'isogénies sur K contient :

1) 11 exemplaires de $X_o(11)$;

2) 5 exemplaires de $J_{ndép}(11)$;

3) 10 exemplaires de $J_{\mathfrak{S}_4}(11)$.

3.4 Considérons maintenant le groupe $H_{dép}$. L'espace de ses invariants est de dimension un dans $\mathfrak{D}_5$.

Soit f une base des invariants de $H_{dép}$ dans $\mathfrak{D}_5$, et soit f' une base des invariants de $H_{ndép}$ dans $\mathfrak{D}_5$. Il existe un élément M de $\mathfrak{G}$ envoyant f' sur f :

$$f = f'|M .$$

On en déduit l'existence d'un diagramme commutatif :

$$\begin{array}{ccc} J(11)_K & \xrightarrow{M} & J(11)_K \\ \downarrow & & \downarrow \\ J_{dép}(11) & \longrightarrow & J_{ndép}(11) \end{array}$$

où le morphisme du bas est surjectif. On montre de même l'existence d'une K-isogénie :

$$J_{dép}(11) \to X_o(11) .$$

3.5 D'autre part, comparons l'expression explicite de la forme $f_6(z)$, donnée par la prop. 1.4.1 du chap. I, avec les séries thêta de 2.2. Il est clair que l'on a :

$$f_6(\frac{z}{11}) = \sum_{j=1}^{5} (\frac{j}{11}).\theta(z,j,\sqrt{-11}).$$

Il en résulte que $J_{ndép}(11)$ est isogène sur K à la courbe E_6. Par conséquent, $J_{dép}(11)$ est K-isogène au produit $X_o(11) \times E_6$, et $J_{\mathfrak{S}_4}(11)$ est isogène sur K à la courbe E_4.

3.6 On a donc retrouvé sous une forme plus faible les résultats du §4 du chap. II, à savoir :

$J_{dép}(11)$ est isogène sur K au produit $X_o(11) \times E_6$;

$J_{ndép}(11)$ est isogène sur K à E_6 ;

$J_{\mathfrak{S}_4}(11)$ est isogène sur K à E_4.

De plus :

Proposition 3.6.1 : La jacobienne $J(11)_K$ du modèle canonique associé aux homothéties (cf. II, ex. 3.7.3) est isogène sur $\mathbb{Q}(\sqrt{-11})$ au produit : $E_1^{11} \times E_6^5 \times E_4^{10}$.

3.7 Une façon plus conceptuelle d'envisager les résultats précédents est la suivante.

3.7.1 Plaçons-nous de nouveau sur le corps K, et considérons le morphisme :

$$G \to \mathrm{Aut}(J(11)_K).$$

Considérons la variété abélienne $J(11)_K$ à K-isogénie près. Par linéarité, on obtient un morphisme :

$$\tilde{\rho} : \mathbb{Q}[G] \to \mathrm{End}(J(11)_K) \otimes \mathbb{Q}.$$

3.7.2 Les résultats de Hecke rappelés en 2.3 signifient que ce morphisme se factorise à travers la projection :

$\mathbb{Q}[G] \to M_{11}(\mathbb{Q}) \times M_5(K) \times M_{10}(\mathbb{Q})$, d'où un morphisme ρ :

$$\rho : M_{11}(\mathbb{Q}) \times M_5(K) \times M_{10}(\mathbb{Q}) \to \mathrm{End}(J(11)_K) \otimes \mathbb{Q}.$$

3.7.3 En particulier, les matrices unités de $M_{11}(\mathbb{Q})$, $M_5(K)$, $M_{10}(\mathbb{Q})$ définissent des projecteurs de dimension 11, 5 et 10 respectivement ; par conséquent, $J(11)_K$ à isogénie près, se coupe en trois morceaux de dimensions 11, 5 et 10, correspondant aux images de ces matrices unités.

3.7.4 Soit $M(i)$ la matrice de $M_{11}(\mathbb{Q})$ dont le terme diagonal i-ième est 1, et dont tous les autres termes sont nuls, $i = 1,\dots,11$. Cette matrice définit un projecteur de dimension 1. D'autre part, deux matrices $M(i)$, $M(j)$ sont conjuguées par une matrice de permutation. On en déduit :

L'image de $\rho(M_{11}(\mathbb{Q}))$ est isogène sur K au produit de 11 courbes elliptiques, isogènes entre elles sur K .

On a un résultat analogue pour $M_5(K)$, $M_{10}(\mathbb{Q})$.

3.7.5 De plus, K est dans le centre de $M_5(K)$, et par conséquent :

L'image de $\rho(M_5(K))$ est isogène sur K au produit de 5 courbes elliptiques dont le corps d'endomorphismes est K .

3.7.6 Soit $\bar{H}$ un sous-groupe de $\mathcal{G}$. Il lui est associé un projecteur, image de l'élément :

$$\frac{1}{|\bar{H}|} \sum_{\bar{h}\in\bar{H}} \bar{h} \quad \text{de} \quad \mathbb{Q}[\mathcal{G}] .$$

L'image de ce projecteur se répartit suivant les trois morceaux correspondant aux représentations $\mathfrak{D}_{11}$, $\mathfrak{D}_5$, $\mathfrak{D}_{10}(3)$. L'image dans le morceau correspondant à $\mathfrak{D}$ est de dimension donnée par la trace de l'élément considéré, soit

$$\frac{1}{|\bar{H}|} \sum_{\bar{h}\in\bar{H}} \chi_{\mathfrak{D}}(\bar{h}), \text{ dimension des invariants de } H.$$

3.7.7 On peut alors procéder comme en 3.3, 3.4 ; en prenant des groupes H convenables, on obtient de nouveau le fait que $J(11)_K$ est isogène au produit $E_1^{11} \times E_6^5 \times E_4^{10}$, et que $J_{\text{dép}}(11)$, $J_{\text{ndép}}(11)$ et $J_{\mathfrak{S}_4}(11)$ sont isogènes sur K à $E_1 \times E_6$, E_6 et E_4 respectivement.

Table 1

Le groupe $\mathcal{H}(121)$ (cf.I.3.2.3)

Cette table donne l'expression des 110 éléments $(\widetilde{c}:\widetilde{1})$, $c = 1,\dots,120$, $(c,11) = 1$, en fonction des douze éléments $(\widetilde{c}:\widetilde{1})$, $c \in \mathcal{B}$, où $\mathcal{B} = \{2,3,4,5,7,8,9,18,26,36,49,51\}$.

On note simplement (c) l'élément $(\widetilde{c}:\widetilde{1})$.

(1) = 0	(45) = 0
(6) = (5)	(46) = -(5) + (7) - (18)
(10) = 0	(47) = -(18)
(12) = 0	(48) = (7) - (8) - (18)
(13) = (9)	(50) = (49)
(14) = -(8) + (9)	(52) = -(2) + (3) - (4) + (7) - (18) - (49) + (51)
(15) = -(8)	(53) = -(2) + (3) - (5) + (7) - (18) - (49) + (51)
(16) = (7) - (8)	(54) = 0
(17) = (7) - (8)	(56) = 0
(19) = -(5) + (7)	(57) = -(2) + (3) - (5) + (7) - (18) - (49) + (51)
(20) = -(5)	(58) = -(2) + (3) - (5) + (7) - (18)
(21) = 0	(59) = -(2) + (3)
(23) = 0	(60) = -(2)
(24) = -(5)	(61) = -(2)
(25) = (4) - (5)	(62) = -(7) + (8) + (18) + (49)
(27) = -(4) + (5) + (26)	(63) = -(7) + (8) + (18)
(28) = -(4) + (5) + (26)	(64) = -(7) + (8)
(29) = -(4) + (5)	(65) = 0
(30) = -(4)	(67) = 0
(31) = (3) - (4)	(68) = -(7) + (8)
(32) = 0	(69) = -(7)
(34) = 0	(70) = (5) - (7)
(35) = (3) - (4)	(71) = (5) - (7) + (18)
(37) = -(3) + (4) + (36)	(72) = (5) - (7) + (18)
(38) = -(3) + (4)	(73) = (2) - (3) + (5) - (7) + (18)
(39) = -(3) + (4)	(74) = -(49) + (51)
(40) = -(3)	(75) = -(49)
(41) = (2) - (3)	(76) = 0
(42) = -(5) + (7) - (18)	(78) = 0
(43) = 0	(79) = -(49)

(80) = (7) - (8) - (18) - (49)
(81) = -(2) + (7) - (8) - (18) - (49)
(82) = -(3) + (4)
(83) = -(3) + (4)
(84) = -(36)
(85) = (3) - (4) - (36)
(86) = (3) - (4)
(87) = 0
(89) = 0
(90) = -(3) + (4)
(91) = -(2) + (3) - (4) + (7) - (8) - (18) - (49)
(92) = -(8)
(93) = -(9)
(94) = -(9)
(95) = (8) - (9)
(96) = (8)
(97) = -(2) + (3) - (4) + (7) - (18) - (49)
(98) = 0
(100)= 0
(101)= -(2) + (3) - (4) + (7) - (18) - (49)
(102)= -(51)
(103)= (49) - (51)
(104)= (2) - (3) + (5) - (7) + (18) + (49) - (51)
(105)= (2) - (3) + (5) - (7) + (18) + (49) - (51)
(106)= -(4) + (5)
(107)= -(26)
(108)= (4) - (5) - (26)
(109)= 0
(111)= 0
(112)= (4) - (5) - (26)
(113)= (4) - (5)
(114)= (2) - (3) + (4) - (7) + (18) + (49) - (51)
(115)= (2) - (3) + (4) - (7) + (18) + (49)
(116)= (2) - (3) + (4) - (7) + (18) + (49)
(117)= (2) - (3) + (4) - (7) + (8) + (18) + (49)
(118)= (2) - (7) + (8) + (18) + (49)
(119)= (2)
(120)= 0

Table 2

Formules de conjugaison complexe

Cette table donne l'expression des transformés des vecteurs $(\tilde{c}:1)$, $c \in \mathcal{B}$, par la conjugaison complexe (cf. I.3.2.5).

On note simplement $(\tilde{c}:\tilde{1}) = (c)$, comme dans la table 1.

$$\overline{(2)} = (2)$$
$$\overline{(3)} = (2) - (7) + (8) + (18) + (49)$$
$$\overline{(4)} = (2) - (3) + (4) - (7) + (8) + (18) + (49)$$
$$\overline{(5)} = (2) - (3) + (4) - (7) + (18) + (49)$$
$$\overline{(7)} = (2) - (3) + (4) - (7) + (18) + (49) - (51)$$
$$\overline{(8)} = (4) - (5)$$
$$\overline{(9)} = (4) - (5) - (26)$$
$$\overline{(18)} = (49) - (51)$$
$$\overline{(26)} = (8) - (9)$$
$$\overline{(36)} = (3) - (4) - (36)$$
$$\overline{(49)} = (5) - (7) + (18)$$
$$\overline{(51)} = (5) - (7)$$

Table 3

Expression des vecteurs $\{0,\frac{1}{c}\}$, $c \in \mathcal{B}\ldots$,

en fonction des vecteurs $\gamma_k^{\pm}$, $k = 1,\ldots,6$

$$\{0,\tfrac{1}{2}\} = \gamma_1^+ - \gamma_2^+ - \gamma_3^+ - \gamma_4^+ - \gamma_6^+$$

$$\{0,\tfrac{1}{3}\} = \tfrac{1}{2}\gamma_1^+ + \tfrac{1}{2}\gamma_1^-$$

$$\{0,\tfrac{1}{4}\} = \tfrac{1}{2}\gamma_1^+ - \gamma_2^+ + \tfrac{1}{2}\gamma_1^-$$

$$\{0,\tfrac{1}{5}\} = \tfrac{1}{2}\gamma_1^+ - \gamma_2^+ - \tfrac{1}{2}\gamma_3^+ + \tfrac{1}{2}\gamma_3^-$$

$$\{0,\tfrac{1}{7}\} = \tfrac{1}{2}\gamma_1^+ - \gamma_2^+ - \tfrac{1}{2}\gamma_3^+ - \tfrac{1}{2}\gamma_4^+ + \tfrac{1}{2}\gamma_4^-$$

$$\{0,\tfrac{1}{8}\} = \tfrac{1}{2}\gamma_3^+ - \tfrac{1}{2}\gamma_1^- + \tfrac{1}{2}\gamma_3^-$$

$$\{0,\tfrac{1}{9}\} = \tfrac{1}{2}\gamma_3^+ - \tfrac{1}{2}\gamma_5^+ + \tfrac{1}{2}\gamma_2^-$$

$$\{0,\tfrac{1}{18}\} = \tfrac{1}{2}\gamma_6^+ - \tfrac{1}{2}\gamma_3^- + \tfrac{1}{2}\gamma_4^- - \tfrac{1}{2}\gamma_5^-$$

$$\{0,\tfrac{1}{26}\} = \tfrac{1}{2}\gamma_5^+ + \tfrac{1}{2}\gamma_1^- + \tfrac{1}{2}\gamma_2^- - \tfrac{1}{2}\gamma_3^-$$

$$\{0,\tfrac{1}{36}\} = \tfrac{1}{2}\gamma_2^+ + \tfrac{1}{2}\gamma_6^-$$

$$\{0,\tfrac{1}{49}\} = \tfrac{1}{2}\gamma_4^+ + \tfrac{1}{2}\gamma_6^+ + \tfrac{1}{2}\gamma_5^-$$

$$\{0,\tfrac{1}{51}\} = \tfrac{1}{2}\gamma_4^+ - \tfrac{1}{2}\gamma_3^- + \tfrac{1}{2}\gamma_4^-$$

Table 4

<u>Coordonnées des vecteurs</u> γ_k^+ , $k = 1,\dots,6$

<u>en fonction des périodes</u> ω_i^+ ,

<u>des réseaux</u> L_i^+ $(i = 1,\dots,6)$ (cf. I.4.9).

<u>Notations</u>.

$$\omega_1^+ = \{0,\tfrac{1}{3}\}_{f_1} + \{0,-\tfrac{1}{3}\}_{f_1} ,$$

$$\omega_2^+ = 5\omega_1^+ ,$$

$$\omega_i^+ = \gamma_{4,i}^+ = \{0,\tfrac{1}{51}\} + \{0,-\tfrac{1}{51}\}_{f_i}$$

$(i = 3,4,5,6)$.

$\gamma_{1,1}^+ = \omega_1^+$	$\gamma_{1,2}^+ = 0$	$\gamma_{1,3}^+ = 10\,\omega_3^+$	$\gamma_{1,4}^+ = -2\omega_4^+$	$\gamma_{1,5}^+ = -2\omega_5^+$	$\gamma_{1,6}^+ = 0$
$\gamma_{2,1}^+ = \omega_1^+$	$\gamma_{2,2}^+ = 0$	$\gamma_{2,3}^+ = 2\,\omega_3^+$	$\gamma_{2,4}^+ = \omega_4^+$	$\gamma_{2,5}^+ = -\omega_5^+$	$\gamma_{2,6}^+ = 0$
$\gamma_{3,1}^+ = \omega_1^+$	$\gamma_{3,2}^+ = 0$	$\gamma_{3,3}^+ = 2\,\omega_3^+$	$\gamma_{3,4}^+ = -2\omega_4^+$	$\gamma_{3,5}^+ = 2\omega_5^+$	$\gamma_{3,6}^+ = 0$
$\gamma_{4,1}^+ = -\omega_1^+$	$\gamma_{4,2}^+ = \omega_2^+$	$\gamma_{4,3}^+ = \omega_3^+$	$\gamma_{4,4}^+ = \omega_4^+$	$\gamma_{4,5}^+ = \omega_5^+$	$\gamma_{4,6}^+ = \omega_6^+$
$\gamma_{5,1}^+ = -\omega_1^+$	$\gamma_{5,2}^+ = -\omega_2^+$	$\gamma_{5,3}^+ = -\omega_3^+$	$\gamma_{5,4}^+ = 0$	$\gamma_{5,5}^+ = 2\omega_5^+$	$\gamma_{6,6}^+ = \omega_6^+$
$\gamma_{6,1}^+ = -\omega_1^+$	$\gamma_{6,2}^+ = -\omega_2^+$	$\gamma_{6,3}^+ = 3\omega_3^+$	$\gamma_{6,4}^+ = 0$	$\gamma_{6,5}^+ = 0$	$\gamma_{6,6}^+ = -\omega_6^+$

Table 5

Coordonnées des vecteurs γ_k^-, $k=1,\dots,6$

en fonction des périodes ω_i^-

des réseaux L_i^- $(i=1,\dots,6)$ (cf. I.4.9).

Notations.

$$\omega_1^- = \{0,\tfrac{1}{3}\}_{f_1} - \{0,-\tfrac{1}{3}\}_{f_1} ,$$

$$\omega_2^- = \omega_1^- ,$$

$$\omega_3^- = -\tfrac{1}{2}\left[\{0,\tfrac{1}{49}\}_{f_3} - \{0,-\tfrac{1}{49}\}_{f_3}\right] ,$$

$$\omega_4^- = \{0,\tfrac{1}{49}\}_{f_4} - \{0,-\tfrac{1}{49}\}_{f_4} ,$$

$$\omega_5^- = \{0,\tfrac{1}{49}\}_{f_5} - \{0,-\tfrac{1}{49}\}_{f_5} ,$$

$$\omega_6^- = \tfrac{1}{2}\left[\{0,\tfrac{1}{49}\}_{f_6} - \{0,-\tfrac{1}{49}\}_{f_6}\right] .$$

$\gamma_{1,1}^- = \omega_1^-$	$\gamma_{1,2}^- = 2\omega_1^-$	$\gamma_{1,3}^- = 0$	$\gamma_{1,4}^- = 0$	$\gamma_{1,5}^- = 0$	$\gamma_{1,6}^- = -2\omega_6^-$
$\gamma_{2,1}^- = 0$	$\gamma_{2,2}^- = -\omega_1^-$	$\gamma_{2,3}^- = 3\omega_3^-$	$\gamma_{2,4}^- = 2\omega_4^-$	$\gamma_{2,5}^- = 0$	$\gamma_{2,6}^- = \omega_6^-$
$\gamma_{3,1}^- = 0$	$\gamma_{3,2}^- = 0$	$\gamma_{3,3}^- = 0$	$\gamma_{3,4}^- = 0$	$\gamma_{3,5}^- = 0$	$\gamma_{3,6}^- = -4\omega_6^-$
$\gamma_{4,1}^- = -\omega_1^-$	$\gamma_{4,2}^- = -\omega_1^-$	$\gamma_{4,3}^- = \omega_3^-$	$\gamma_{4,4}^- = \omega_4^-$	$\gamma_{4,5}^- = \omega_5^-$	$\gamma_{4,6}^- = -\omega_6^-$
$\gamma_{5,1}^- = 0$	$\gamma_{5,2}^- = -2\omega_1^-$	$\gamma_{5,3}^- = -2\omega_3^-$	$\gamma_{5,4}^- = \omega_4^-$	$\gamma_{5,5}^- = \omega_5^-$	$\gamma_{5,6}^- = 2\omega_6^-$
$\gamma_{6,1}^- = \omega_1^-$	$\gamma_{6,2}^- = -2\omega_1^-$	$\gamma_{6,3}^- = 4\omega_3^-$	$\gamma_{6,4}^- = -\omega_4^-$	$\gamma_{6,5}^- = \omega_5^-$	$\gamma_{6,6}^- = 2\omega_6^-$

Table 6

Soit $z = \frac{1}{2} \sum_{k=1}^{6} (a_k \gamma_k^+ + b_k \gamma_k^-)$, $a_k, b_k \in \mathbb{R}$, un vecteur arbitraire de C^6 .

Soit $\pi_i(z) = s_i w_i^+ + t_i w_i^-$, $i = 1,\ldots,6$, la projection de z sur la $i^{\text{ième}}$ coordonnée (cf. I.4.9).

On a

$$12.\begin{pmatrix} a_1 \\ \vdots \\ a_6 \end{pmatrix} = M^+ . \begin{pmatrix} s_1 \\ \vdots \\ s_6 \end{pmatrix} ,$$

$$12.\begin{pmatrix} b_1 \\ \vdots \\ b_6 \end{pmatrix} = M^- . \begin{pmatrix} t_1 \\ \vdots \\ t_6 \end{pmatrix} , \text{ avec :}$$

$$M^+ = \begin{bmatrix} -2 & -1 & 1 & -4 & -4 & 6 \\ 12 & -6 & 2 & 12 & 4 & 0 \\ 6 & 3 & 1 & 0 & 8 & -6 \\ -4 & 10 & 2 & 4 & 4 & 0 \\ 0 & -12 & 0 & 0 & 0 & 12 \\ -4 & -2 & 2 & 4 & 4 & -12 \end{bmatrix}$$

$$M^- = \begin{bmatrix} 8 & 8 & 0 & 4 & 12 & 0 \\ 4 & -2 & 2 & 8 & -8 & 0 \\ 6 & -9 & -1 & 0 & -8 & -6 \\ -12 & 6 & 2 & 0 & 16 & 0 \\ 8 & -4 & -4 & 4 & 4 & 0 \\ 4 & -2 & 2 & -4 & 4 & 0 \end{bmatrix}$$

Table 7

Pointes de $X_o(121)$

Soit u un entier. On pose (cf. I, 4.10.2) :

$$2\{i\infty,\tfrac{u}{11}\}_{f_i} = X(u,i)\omega_i^+ + Y(u,i)\omega_i^- \ , \ i=1,\dots,6 \ .$$

Cette table donne les valeurs de $(X(u,i),Y(u,i))$ pour $u = 0,1,3,4,5,9$, $i=1,\dots,6$.

i \ u	0	1	3	4	5	9
1	$-\frac{2}{5},0$	$0,0$	$-1,1$	$1,1$	$2,0$	$-2,0$
2	$0,0$	$\frac{2}{25},0$	$-\frac{3}{25},1$	$\frac{7}{25},1$	$\frac{12}{25},0$	$-\frac{8}{25},0$
3	$-4,0$	$-2,\frac{32}{25}$	$1,\frac{27}{25}$	$3,-\frac{13}{25}$	$0,-\frac{8}{25}$	$0,-\frac{28}{25}$
4	$2,0$	$1,\frac{9}{11}$	$-1,\frac{1}{11}$	$0,\frac{4}{11}$	$-1,\frac{3}{11}$	$0,-\frac{6}{11}$
5	$2,0$	$1,\frac{1}{11}$	$1,\frac{3}{11}$	$0,\frac{4}{11}$	$-3,\frac{5}{11}$	$0,-\frac{2}{11}$
6	$0,0$	$-\frac{6}{11},0$	$\frac{1}{11},-1$	$\frac{3}{11},-1$	$\frac{4}{11},0$	$-\frac{2}{11},2$

INDEX

BIBLIOGRAPHIE

[1] A. ATKIN, J. LEHNER.- Hecke operators on $\Gamma_o(m)$, Math. Ann., 185, (1970), p. 134-160.

[2] B.J. BIRCH, W. KUYK (ed.).- Numerical tables on elliptic curves, in Modular Functions of One Variable IV, Lecture Notes in Math. 476, p. 74-144, Berlin-Heidelberg-New York, Springer (1975).

[3] E. HECKE.- Ueber ein Fundamentalproblem aus der Theorie der elliptischen Modulfunktionen, Abh. Math. Sem. Hamburg 6 (1928), p. 235-257 [Mathematische Werke, Göttingen, Vandenhoeck and Ruprecht (1959), p. 525-547].

[4] E. HECKE.- Grundlagen einer Theorie der Integralgruppen und der Integralperioden bei den Normalteilern der Modulgruppe, Math. Ann. 116 (1939), p. 469-510 [Mathematische Werke, Göttingen, Vandenhoeck and Ruprecht (1959), p. 731-772].

[5] D. KUBERT, S. LANG.- Units in the Modular Function Field. II. A full Set of Units, Math. Ann. 218 (1975), p. 175-189.

[6] J. LEHNER.- A short Course in Automorphic Functions, Holt, Rinehart and Winston (1966).

[7] G. LIGOZAT.- Courbes modulaires de genre 1, Bull. Soc. Math. France, Suppl., Mém. N° 43, 80p (1975).

[8] G. LIGOZAT.- Groupes cuspidaux des courbes modulaires de niveau p, (à paraître).

[9] Y. MANIN.- Points paraboliques et fonctions zêta des courbes modulaires (en russe), Izv. Akad. N. S.S.S.R., 36 (1972), p. 19-66 [trad. anglaise : Math. USSR-Izvestja, 6 (1972), p. 19-64].

[10] B. MAZUR.- Courbes elliptiques et symboles modulaires, Sém. Bourbaki, 24e année (1971/72), exposé n° 414, Lecture Notes in Math., 317, Berlin-Heidelberg-New York, Springer (1973).

[11] A. OGG.- Rational points on certain elliptic modular curves, Proc. Symp. Pure Math., vol. 24, Amer. Math. Soc., Providence R.I. (1973), p. 221-231.

[12] J.-P. SERRE.- Représentations linéaires des groupes finis (2ème édition), Paris, Hermann (1971).

[13] G. SHIMURA.- Introduction to the arithmetic theory of automorphic functions, Publ. Math. Soc. Japan, N° 11, Iwanami Shoten and Princeton University Press (1971).

[14] G. SHIMURA.- On the factors of the jacobian variety of a modular function field, J. Math. Soc. Japan, vol. 25, n° 3 (1973), p. 523-544.

[15] G. SHIMURA.- On elliptic curves with complex multiplication as factors of the jacobians of modular function fields, Nagoya Math. J., vol. 43 (1971), p. 199-208.

[16] J. VÉLU.- Courbes elliptiques sur $\mathbb{Q}$ ayant bonne réduction en dehors de $\{11\}$, C. R. Acad. Sci. Paris, 273 (1971), p. 73-75.

G. Ligozat
Université de Paris-Sud
Centre d'Orsay
Mathematiques, Bâtiment 425
91405 Orsay, France

ON THE NORMALIZER OF $\Gamma_0(N)$.

by P.G. Kluit

0. INTRODUCTION.

Let N be a positive integer and let $\Gamma_0(N)$ be the subgroup of the modular group $\Gamma = SL(2,\mathbb{Z})/(\pm 1)$ defined by the matrices $\begin{pmatrix} a & b \\ c & d \end{pmatrix}$ with N dividing c.

Let H as usual be the upper half plane and $H^* = H \cup \mathbf{Q} \cup \{\infty\}$, then the quotient $X_0(N) = H^*/\Gamma_0(N)$ is a compact Riemann surface. Let $\Gamma^*(N)$ be the subgroup of $SL(2,\mathbf{R})/\{\pm 1\}$ consisting of the matrices

$$W_m = m^{-1/2}\begin{pmatrix} a & b \\ c & d \end{pmatrix} \quad \text{with } m|N,\ N|c,\ m|a \text{ and } m|d;\ a,b,c,d \in \mathbf{Z}.$$

(This implies $\frac{a}{m}\frac{d}{m}m - b\frac{c}{N}\frac{N}{m} = 1$, so $(m, N m^{-1}) = 1$). Then $\Gamma^*(N)$ is the abelian normalizer of $\Gamma_0(N)$ in $SL(2,\mathbf{R})/\{\pm 1\}$. In case neither 4 nor 9 divides N it is even the <u>full</u> normalizer. The factor group $G(N) = \Gamma^*(N)/\Gamma_0(N)$ is abelian of type $(2,2,\dots,2)$, the number of direct factors being equal to the number of prime factors dividing N. We may conceive this group as a subgroup of the automorphism group of $X_0(N)$; its elements are the involutions w_m of Atkin-Lehner type. (Here w_m denotes the image of some W_m in $G(N)$). See for all this : [7], [1] and [8].

1. THE FIXED POINTS OF W_m.

Take $N = mm'$, with $(m,m') = 1$. For any such N and m $(\neq 1)$ we will calculate the number $\nu(N,m)$ of fixed points of the involution w_m on $X_0(N)$, i.e. the number of ramification points of the quotient map :

$$X_0(N) \xrightarrow{\ \phi_{m,N}\ } X_0(N)/(w_m)$$

Observe that the degree of $\phi_{m,n}$, and hence the ramification-index, is 2.

In case $N = m$ we have Fricke's result [4]; for $m \geq 5$ it reads :

$$\nu(m,m) = \begin{cases} h(-4m) & \text{if } m \not\equiv -1 \bmod 4 \\ h(-4m) + h(-m) & \text{if } m \equiv -1 \bmod 4 . \end{cases}$$

Starting from this we may obtain a formula for $\nu(N,m)$, using the following diagrams :

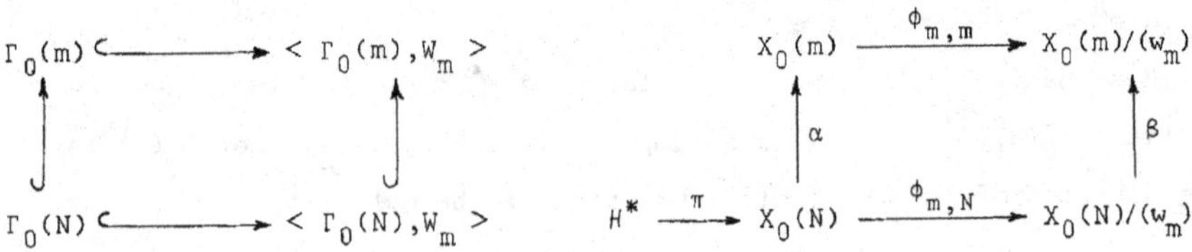

Observe that : $\Gamma_0(m) \cap < \Gamma_0(N), W_m > = \Gamma_0(N)$. For W_m we may take the same matrix above and below; the w_m have a different meaning each time, but this will give no confusion.

If $\phi_{m,N}$ ramifies in $a \in X_0(N)$, say $a = \pi(z)$, then $\exists\ M \in W_m.\Gamma_0(N)$ such that $Mz = z$. Then evidently $M \in W_m.\Gamma_0(m)$, so $\phi_{m,m}$ ramifies in $\alpha(a)$. Conversely, let $b \in X_0(m)$ be a point of ramification of ϕ; say $b = \alpha \circ \pi\ (z)\ (z \in H^*)$. Then $\exists\ M \in W_m.\Gamma_0(m)$ such that $Mz = z$. Let $\{R_i\}$ be a set of coset-representatives of $\Gamma_0(N)$ in $\Gamma_0(m)$. Then $\alpha^{-1}(b) = \{\pi \circ R_i(z)\}$. Any of these points is a ramification point of $\phi_{m,N}$ if and only if $\exists\, T \in W_m.\Gamma_0(N)$ such that $R_i^{-1}\, T\, R_i(z) = z$. Now if $m \geq 5$ then $W_m.\Gamma_0(m)$ contains no parabolic elements, and elliptic elements only of order 2. It follows that $\pi \circ R_i(z)$ is a point of ramification of $\phi_{m,N}$ precisely if $R_i\, M R_i^{-1} \in W_m.\Gamma_0(N)$.

As we can explicitly give a set $\{R_i\}$ of representatives (cf. [10] p. 25), this last condition may be written out for any M, and reduces to some congruence mod m'.

Calculations can be simplified by doing these things step by step; first for the powers of 2 (the prime 2, as usual, needs special attention; this time because of the occurrence of 4m in Fricke's formula). So the following diagram is used, first for $N = m$, and $q = 2^{\lambda}$; then for N arbitrarily, q an odd prime :

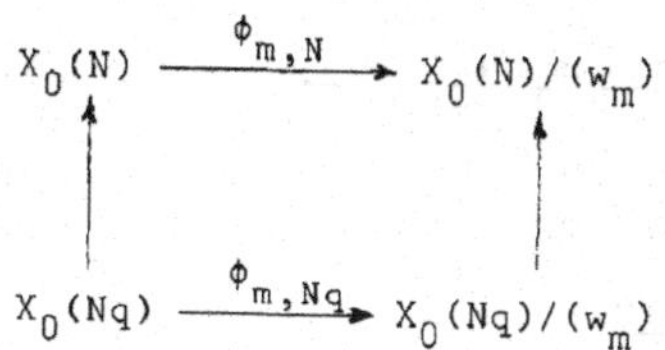

If $m = 2,3$ or 4 the cosets $W_m . \Gamma_0(m)$ contain elliptic elements of order 4 and 6 resp. parabolic elements. It may happen then, that α is ramified in $a \in X_0(N)$, and that $\phi_{m,m}$ is ramified in $\alpha(a)$, so these cases need special attention. We will not go into details now, but just state the results.

<u>THEOREM.</u>

0) <u>If</u> $m \geq 5$: $\nu(m,m) = h(-4m)$ <u>if</u> $m \not\equiv -1 \bmod 4$

$h(-4m) + h(-m)$ <u>if</u> $m \equiv -1 \bmod 4$.

1) <u>If</u> $m \geq 5$ <u>and odd</u> :

$$\nu(m2^{\lambda},m) = \begin{cases} h(-4m) & \lambda = 1 \\ 0 & \lambda > 1 \end{cases} \quad m \equiv 1 \bmod 4$$

$$\begin{cases} h(-4m) + 3h(-m) & \lambda = 1 \\ 2h(-4m) + 2(1 + (\frac{-m}{2}))h(-m) & \lambda = 2 \\ 2(1 + (\frac{-m}{2}))\nu(m,m) & \lambda > 2 \end{cases} \quad m \equiv -1 \bmod 4$$

2) <u>If</u> $m \geq 5$, m' <u>odd and</u> $(m,2^{\lambda}m') = 1$ $(\lambda = 0,1,2,\ldots..)$

$$\nu(mm'2^{\lambda},m) = \prod_{p|m'} (1 + (\frac{-m}{p}))\ \nu(m2^{\lambda},m)$$

3) (i) *If* m' *odd* : $\nu(2m',2) = \prod_{p|m'} (1 + (\frac{-1}{p})) + \prod_{p|m'} (1 + (\frac{-2}{p}))$

(ii) *If* $(m',3.2^{\lambda}) = 1$: $\nu(3.2^{\lambda}.m',3) = 2 \prod_{p|m'} (1 + (\frac{-3}{p}))$ *if* $\lambda = 0,1,2$

$= 0$ *if* $\lambda > 2$

(iii) *If* m' *odd* : $\nu(4m';4) = \prod_{p|m'} (1 + (\frac{-1}{p})) + \sum_{d|m'} \phi((d,m'd^{-1}))$

(*Here* ϕ *denotes Euler's function*).

2. WEIERSTRASS POINTS.

As is indicated in [7] these formulae may be helpful to determine Weierstrass points of $X_0(N)$. We have Schoeneberg's theorem:

THEOREM. *Let* τ *be a fixed point of an automorphism* w *of* X *of period* $p > 1$; *let* g *be the genus of* X, g^* *the genus of* $X/\langle w \rangle$. *If* $g^* \neq [\frac{g}{p}]$, *then* τ *is a Weierstrass point of* X.

Applying this to the case $w = w_m$, $X = X_0(N)$, $p = 2$, and using Hurwitz' formula :

$$g - 2g^* = -1 + \frac{1}{2}\nu(N,m),$$

we find by simple calculation :

COROLLARY. *If* $\nu(N,m) > 4$ *then the fixed points of* w_m *are Weierstrass points*.

The cases in which $0 < \nu(N,m) < 4$ may be determined more or less explicitly. We make the following observations :

a) Assume $N \neq m$, $\nu(m,m) = 2$. It is known that $h(d) \leqslant 2 \Rightarrow d \geqslant -427$, so we have explicitly : $\nu(m,m) = 2 \Rightarrow m \leqslant 10$ or $m = 12,13,16,18,22,25,28,37,58$.

Then $\nu(N,m)$ is either zero or > 4, provided that Nm^{-1} is sufficiently composite (this can be made specific in each case; for $m \geq 5$ two odd primes is sufficient).

b) Assume $N \neq m$, $\nu(m,m) > 2$. Then always : $\nu(N,m) = 0$ or $\nu(N,m) > 4$.

c) Assume $N = m$. This is the case handled by Lehner and Newman [7]. They proved : $\nu(m,m) \leq 4$ for only finitely many m. If $m \not\equiv -1 \bmod 4$ we have $\nu(m,m) = h(-4m)$. This reduces the problem to the rather hard question of determining all m for which $h(-4m) = 4$. By machine program I found : $\nu(m,m) > 4$ if $252 < m < 4000$. It seems reasonable to expect $\nu(m,m) > 4$ if $m > 4000$ as well.

There is at least one instance (i.e. $N = m = 37$), for which $\nu(N,m) = 2$, and the fixed points of w_m are certainly not Weierstrass points (see [8]). On the other hand, for almost all N this method gives some Weierstrass points of $X_0(N)$. Unfortunately, in only very few cases it gives <u>all</u> the Weierstrass points.

3. GENERA AND INVARIANTS.

Let G be any subgroup of G(N) of order 2^α, let g be the genus of $X_0(N)$, and let g^* be the genus of the quotient-curve : $X_0(N)/G$. Then applying Hurwitz' formula again we have :

$$g^* = 1 + 2^{-\alpha}\left(g - 1 - 1/2 \sum_{w_m \in G} \nu(N,m)\right).$$

This is obvious once we realize that a point x of $X_0(N)$ is a ramification point of the map $X_0(N) \to X_0(N)/G$ if and only if it is a fixed point of some $w_m \in G$. A common fixed point of w_{m_1} and w_{m_2} would have a stabilizer group cyclic of order 4 in G, which is impossible.

For the case $G = G(N)$, with N squarefree, this formula was given by H. Helling [5], be it in a slightly different form.

A special case, which had much attention already, is given by the groups $\Gamma_0^*(N) = < \Gamma_0(N), W_N >$. Let us denote the genus of the quotient curve $X_0(N)/(w_N) = X_0^*(N)$ by $g_1(N)$. Fricke calculated $g_1(N)$ up to 71, (making a mistake for N = 59) [4].
Helling [6] claims that $g_1(N) \leq 1 \Rightarrow N \leq 131$, referring to existing tables [3] for N < 1000, and giving an estimation for $N \geq 1000$. His estimation however is easily seen to be valid only for $N \geq 3600$. Calculating $g_1(N)$ up to N = 4000 by computer program I verified that indeed $g_1(N) \leq 1 \Rightarrow N \leq 131$.

A more elegant method to find all solutions of $g_1(N) = 0$ is obtained using a result of Ogg [8]. If $X_0^*(N)$ is rational, then the map $X_0(N) \to X_0^*(N)$ gives rise to a function of degree 2 on $X_0(N)$. So in that case $X_0(N)$ must be either rational, elliptic or hyperelliptic. Now the main result of [8] is a complete list of all N for which $X_0(N)$ is hyperelliptic. The cases for which $X_0(N)$ has genus 0 or 1 are easily determined (they are given in [8] as well). This reduces the problem to the range $N \leq 71$, where even Fricke's table is sufficient. (By an extension of Ogg's method of estimation in [8], theorem 3, we can find a manageable bound for $g_1(N) \leq 2$. This will be worked out elsewhere).

Let us now restrict ourselves to the case $g_1(N) = 0$. For any such N there exists a function $j_N : H \to \mathbf{C}$, fulfilling the following conditions :

1° j_N is holomorphic on H

2° j_N is invariant under $\Gamma_0^*(N)$

3° As a function on $X_0^*(N)$, $j_N(z)$ is injective, and has a simple pole at $i\infty$.

This last condition says, that $j_N(z)$ has a Fourier series :

$$j_N(z) = \sum_{n=-1}^{\infty} c(n)\exp(2\pi i n z).$$

The function can be normalized such that $c(-1) = 1$; $c(0) = 0$. With this normalization it is unique. For $N = 2,3$ these functions have been known a long time; they are discussed in [9] and [11]. In [2] B.J. Birch constructs j_{50}, using η-functions. Following this and other methods I constructed j_N for all other N for which $g_1(N) = 0$, and found as a remarkable result :

LEMMA. Under the above conditions we have : $c_N(n) \in \mathbf{Z} \quad \forall n \in \mathbf{N}$.

(Of course we easily obtain : $c_N(n) \in \mathbf{Q}$, for all n, with bounded denominators, by writing j_N as a quotient of modular forms with integral coefficients). I expect the same thing to happen with any subgroup $G \subset G(N)$ giving a rational quotient curve $X_0(N)/G$.

The method of η-functions was not always successful in constructing j_N; sometimes θ-series and Eisenstein series had to be used. In other cases however more than one solution was found, giving rise to remarkable identities. We conclude with an example of these :

$$j_{36}(z) = \frac{\eta^8(6z).\eta(4z).\eta(9z)}{(\eta(2z).\eta(18z))^2.(\eta(3z).\eta(12z))^3}$$

$$= \frac{\eta(9z).\eta(4z)}{\eta(z).\eta(36z)} - 1$$

$$= \frac{(\eta(2z).\eta(18z))^3.\eta(3z).\eta(12z)}{(\eta(z).\eta(36z))^2.\eta(4z).\eta(9z).\eta^2(6z)} - 2.$$

REFERENCES

[1] ATKIN A.O.L., LEHNER J. : Hecke operators on $\Gamma_0(m)$, Math. Ann. 185, pp. 134-160 (1970).

[2] BIRCH B.J. : Some calculations of modular relations, Modular Functions of One Variable I, Springer-Verlag, Lecture Notes in Math. 320, pp. 175-186.

[3] FELL H., NEWMAN M., ORDMAN E. : Tables of genera of groups of linear fractional transformations, Journal of Research of the National Bureau of Standards, P. Math. and Math. Physics 678 (1), (1963)

[4] FRICKE R. : Lehrbuch der Algebra, vol. 3, Braunschweig, Vieweg (1928)

[5] HELLING H. : On the commensurability class of the rational modular group, Journal of the London Math. Soc. (2), pp. 67-72 (1970)

[6] : Note über das Geschlecht gewisser arithmetischer Gruppen, Math. Ann. 205, pp. 173-179 (1973).

[7] LEHNER J. and NEWMAN M. : Weierstrass points of $\Gamma_0(n)$, Annals of Math., vol. 79 (2) (1964).

[8] OGG A.P. : Hyperelliptic modular curves, Bull. Soc. Math. France 102, pp. 449-462 (1974)

[9] RALEIGH J. : The Fourier coefficients of the invariants $j(2^{1/2};\tau)$ and $j(3^{1/2};\tau)$, Transactions of the A.M.S., vol. 87 (1958), pp. 90-107.

[10] SHIMURA G. : Introduction to the arithmetic theory of automorphic functions, Princeton University Press (1971)

[11] SMART J.R. : Parametrization of the automorphic forms for the Hecke groups $G(\sqrt{2})$ and $G(\sqrt{3})$, Duke Math. J., vol. 1, nr. 3, pp. 395-404 (1964)

P. Kluit
Vrije Universiteit
de Boelelaan
1081 Amsterdam

UNITS IN THE MODULAR FUNCTION FIELD

by

Dan Kubert* and Serge Lang*

*Supported by NSF grants

We shall mostly summarize what is in the list of papers [KL] of the bibliography, but we also include some new results. We only sketch the ideas of some of the proofs, referring to the original papers for a complete treatment.

§1. UNITS AND DIVISOR CLASSES AT INFINITY

Let H be the upper half plane. Let $H^* = H \cup \mathbb{Q} \cup \infty$. Let $\Gamma(N)$ be the subgroup of $SL_2(\mathbb{Z}) = \Gamma(1)$ consisting of the matrices $\equiv 1 \bmod N$. There is an affine curve $Y(N)$ defined over $\mathbb{Q}(\mu_N)$ such that $Y(N)_{\mathbb{C}} \approx \Gamma(N)\backslash H$, and its complete curve $X(N)$ is such that $X(N)_{\mathbb{C}} \approx \Gamma(N)\backslash H^*$, where the isomorphism is complex analytic. The cusps, or points at infinity, are by definition $\mathbb{Q}$, ∞, or their images in $X(N)$, so the complement of $Y(N)$ in $X(N)$.

We let R_N be the integral closure of $\mathbb{Z}[j]$ in the function field F_N of $X(N)$ over $\mathbb{Q}(\mu_N)$, and $\mathbb{Q}R_N$ the integral closure of $\mathbb{Q}[j]$. Elements of $(\mathbb{Q}R_N)^*$ will be called <u>units</u>, and elements of R_N^* will be called <u>units over $\mathbb{Z}$</u>. The rank of $(\mathbb{Q}R_N^*)$/constants is obviously bounded by No. of cusps - 1.

THEOREM 1.1. <u>The rank of the units mod constants is equal to the number of cusps - 1.</u>

Viewing $X(N)$ as embedded in its Jacobian, this theorem is equivalent to the fact that the cusps are of finite order in the Jacobian. Let x_1, x_2 be two cusps. We denote by $\{x_1,x_2\}$ the functional on the space of differentials of first kind given by

$$\{x_1,x_2\}: \omega \mapsto \int_{x_1}^{x_2} \omega .$$

A priori, $\{x_1,x_2\}$ lies in $H^1(X(N),\mathbb{R})$. Manin and Drinfeld have shown that $\{x_1,x_2\}$ in fact lies in $H^1(X(N),\mathbb{Q})$. This is a third equivalent statement to the above. They use Hecke operators. Their method suggests generalizations to higher dimensional bounded symmetric domains. We shall leave this method aside in the present exposition, but shall discuss the manner in which explicit units may be constructed to prove the theorem.

Units may be characterized as elements of the function field all of whose zeros and poles are at the cusps. Let f be a unit, and let

$$f = \sum a_n q^{n/N}$$

be its q-expansion, which may have a finite number of terms with $n < 0$. Let $a_m \neq 0$ be the lowest non-zero coefficient. We also write $a_m = a_m(f)$. The following criterion will be used constantly.

If the coefficients $a_n(f\circ\gamma)$ *are algebraic integers, and the lowest coefficient* $a_m(f\circ\gamma)$ *is a unit in* $\mathbf{Q}(\mu_N)$ *for all* $\gamma \in SL_2(\mathbf{Z})$, *then* f *is a unit over* $\mathbf{Z}$.

§2. *CONSTRUCTION OF UNITS, KLEIN FORMS*

Modular functions whose only zeros and poles are at the cusps have been constructed previously as in Newman [Ne], Ogg [O], and also [L 1], Chapter 18, §6. Here we follow [KL II]. Let L be a lattice, $L = [\omega_1, \omega_2]$. The Weierstrass zeta function satisfies

$$\zeta(z+\omega, L) = \zeta(z, L) + \eta(\omega, L),$$

where $\eta(z,L)$ is $\mathbf{R}$-linear in z. In fact, it can be shown that

$$\eta(z,L) = s_2(L)z + \frac{\pi}{\mathbf{N}L}\bar{z},$$

where $\mathbf{N}L$ is the area of a fundamental domain, and $s_2(L)$ is the limit $\lim_{s\to 0} \sum_{\omega\neq 0} 1/\omega^2|\omega|^{2s}$. Cf. appendix. We won't need this. We define the *Klein forms* (homogeneous of degree 1 in z, L) by

$$\underline{k}(z,L) = e^{-\eta(z,L)z/2}\sigma(z,L).$$

Let $\psi(\omega, L) = 1$ if $\omega/2 \in L$ and -1 otherwise. Then

$$\underline{k}(z+\omega, L) = \underline{k}(z,L)\, e^{-\frac{\pi i}{\mathbf{N}L}\operatorname{Im}(z\bar{\omega})}\psi(\omega, L).$$

We shall abbreviate $W = \begin{pmatrix}\omega_1\\ \omega_2\end{pmatrix}$ and $W_\tau = \begin{pmatrix}\tau\\ 1\end{pmatrix}$. We write

$$z = a_1\omega_1 + a_2\omega_2 = a\cdot W$$

where $a = (a_1, a_2)$ is a pair of real numbers, not both in $\mathbb{Z}$. We then also put $\underline{k}_a(W) = \underline{k}(z,L)$. For $\lambda \in \mathbb{C}^*$ we have

$$\underline{k}_a(\lambda W) = \lambda \underline{k}_a(W).$$

If $b = (b_1, b_2) \in \mathbb{Z}^2$ then

$$\underline{k}_{a+b}(W) = \varepsilon_o(a,b)\underline{k}_a(W)$$

where

$$\varepsilon_o(a,b) = (-1)^{b_1b_2+b_1+b_2} e^{-2\pi i(b_1a_2-b_2a_1)/2}$$

has absolute value 1. As usual we put $\underline{k}_a(\tau) = \underline{k}_a(W_\tau)$. Then we define the Siegel function (homogeneous of degree 0)

$$\underline{k}_a(\tau)\Delta^{1/12}(\tau) = g_a(\tau)$$

$$= - q_\tau^{\frac{1}{2}B_2(a_1)} e^{2\pi i a_2(a_1-1)/2} (1-q_z) \prod_{n=1}^{\infty} (1-q_\tau^n q_z)(1-q_\tau^n/q_z)$$

where $q_\tau = e^{2\pi i\tau}$ and $q_z = e^{2\pi iz}$, and $B_2(X) = X^2 - X + \frac{1}{6}$.

We note that $z \longmapsto \log|g(z)|$ is the local component of the Néron-Tate height function at infinity, cf. [Ner], last page.

We shall be concerned with $(a_1, a_2) = (r_1/N, r_2/N)$, $r_1, r_2 \in \mathbb{Z}$. We may take $0 \leqq a_1, a_2 \leqq 1$, and it is convenient to write $\langle t \rangle$ for the smallest real number $\geqq 0$ in the residue class of t mod $\mathbb{Z}$. The expression for $\varepsilon_o(a,b)$ above shows that $\varepsilon_o(a,b)^{2N} = 1$. Also if $\alpha = \begin{pmatrix} a & b \\ c & d \end{pmatrix} \in SL_2(\mathbb{Z})$,

$$\underline{k}_a(\alpha W) = \underline{k}_{a\alpha}(W) = \varepsilon(\alpha)\underline{k}_a(W),$$

this last equality holding for $\alpha \in \Gamma(N)$ with $\varepsilon(\alpha) =$

$$-(-1)^{(\frac{a-1}{N}a_1+\frac{c}{N}a_2+1)(\frac{b}{N}a_1+\frac{d-1}{N}a_2+1)} e^{2\pi i(ba_1^2+(d-a)a_1a_2-ca_2^2)/2N^2}.$$

Therefore $\varepsilon(\alpha)^{2N} = 1$, $\varepsilon(\alpha) = 1$ *if* $\alpha \in \Gamma(2N^2)$, $\underline{k}_a$ *is a modular form on* $\Gamma(2N^2)$, *and* $\underline{k}_a^{2N}$ *is on* $\Gamma(N)$.

THEOREM 2.1. *If* $a = r/N$, *write* $\underline{k}_r$ *instead of* $\underline{k}_a$. *A product*

$$\prod_r \underline{k}_r^{m(r)}$$

(with a finite family $m(r)$ *of integers* $\neq 0$) *has level* N *if and only if*

$$\sum m(r)r_1^2 \equiv \sum m(r)r_2^2 \equiv \sum m(r)r_1r_2 \equiv 0 \bmod N$$

if N *is odd, and if* N *is even, then the first two congruences should be* $\equiv 0 \bmod 2N$.

THEOREM 2.2. *Let* $a \in \frac{1}{N}\mathbb{Z}^2/\mathbb{Z}^2$ *be primitive.*

1) *If* N *is composite, then* g_a *is a unit over* $\mathbb{Z}$.
2) *If* $N = p^n$ *is a prime power, then* g_a *is a unit in* $R_N[1/p]$.
3) *If* $c \in \mathbb{Z}$, $c \neq 0$ *is prime to* N, *then* g_{ca}/g_a *is a unit over* $\mathbb{Z}$.

In particular, if we have a point $y \in Y(N)$ in a number

field K such that $j(y) \in \underline{o}_K$, then we get a homomorphism

$$R_N^* \longrightarrow R_N^*(y) \subset \underline{o}_K^*$$

thus parametrizing units in K. These may be called modular units in the field of N-th division points of an elliptic curve over K. Analogously, if $j(y)$ is not integral, one gets by pull back a homomorphism from the divisors at infinity on $X(N)$ into the group of fractional ideals, inducing a homomorphism from divisor classes to ideal classes, of which it would be very interesting to know significant examples when it is non-degenerate. Cf. [KL I], the last chapter of [L], Ramachandra [Ra] and Robert [Ro] for cases with complex multiplication. The Shimura reciprocity law can be used with good effect in this case to describe the behavior of the special units under the Galois group.

In [KL II], the rank of the units over $\mathbb{Z}$ is determined, and a criterion is given for a unit to be a unit over $\mathbb{Z}$ as follows. Let

$$g = \prod_a g_a^{m(a)}$$

be a finite product, with integer exponents $m(a)$, $a \in \mathbb{Q}^2$ and $a \notin \mathbb{Z}^2$. Write

$$g = g_{comp} \prod_p g^{(p)},$$

where g_{comp} is the partial product taken over those a

which have composite period with respect to $\mathbb{Z}^2$, and $g^{(p)}$ is the product taken over those a having p-power period for the prime p. We know that g_{comp} is a unit over $\mathbb{Z}$, and it is easy to see that

g is a unit over $\mathbb{Z}$ if and only if $g^{(p)}$ is a unit over $\mathbb{Z}$ for each prime p.

Now fix a prime p, and suppose

$$g = g^{(p)} = \prod g_a^{m(a)}$$

is expressed as a product where a has p-power period mod $\mathbb{Z}^2$. Using the distribution relation (see Theorem 5.1 below), we can achieve that g is equal to such a product in which the exact periods of elements a are all equal to a fixed power p^n, in which case we say that the product is normalized. Note that $\mathbb{Z}_p^*$ operates on $\mathbb{Q}_p^2/\mathbb{Z}_p^2$. We consider the orbits under this action.

THEOREM 2.3. *Let $g = g^{(p)} = \prod g_a^{m(a)}$ be a normalized product. Then g is a unit over $\mathbb{Z}$ if and only if for each orbit of $\mathbb{Z}_p^*$ we have*

$$\sum_{a \in \text{orbit}} m(a) = 0.$$

We could also formulate a similar criterion without dealing with normalized products.

Robert's construction of units in the complex multiplication case is seen to come from the following modular construction. Let $L \subset L'$ be lattices with $(L':L)$ odd. Define

$$\underline{k}(z,L'/L) = \underline{k}(z,L)^{(L':L)}/\underline{k}(z,L').$$

Then $\underline{k}(z,L'/L)$ is elliptic and periodic with respect to L (the anti-holomorphic term and $\psi(\omega,L)$ drop out!). We have

$$\underline{k}(z,L'/L) = \prod \frac{1}{\wp(z,L) - \wp(w,L)}$$

where the product is taken over $w \in (L'/L)/\pm 1$, and $w \neq 0$, because both sides have the same zeros and poles.

Let $\alpha \in \mathrm{Mat}_2^+(\mathbb{Z})$ be an integral matrix with odd determinant, and let $N\alpha = \det \alpha$. Define

$$\varphi(z,\alpha,W) = \underline{k}(z,W)^{N\alpha}/\underline{k}(z,\frac{\alpha W}{N\alpha}).$$

In view of the periodicity property of $\underline{k}(z,L'/L)$, the above value depends only on the left coset of α with respect to $SL_2(\mathbb{Z})$, so we may assume that

$$\alpha = \begin{pmatrix} a & b \\ 0 & d \end{pmatrix}$$

is triangular. Let $A = \{\alpha_j, n_j\}_{j \in J}$ be a finite family, where α_j are matrices as above, and n_j are integers such that $\sum n_j(\underset{\sim}{N}\alpha_j - 1) = 0$. We let

$$\varphi(z,A,W) = \prod \varphi(z,\alpha_j,W)^{n_j}.$$

Then φ has weight 0 in (z,W), and is an even elliptic function. For $z = a\cdot W$ its q-expansion is easily determined from that of the Klein forms. We assume again that $a = (r_1/N, r_2/N)$ has denominator N, and similarly $b = (b_1,b_2)$. We let

$$\varphi(a\cdot W,\alpha,W) = \varphi_a(\alpha,W),$$

and similarly for $\varphi_a(A,W)$, replacing α by α_j and taking the product raising each term to the power n_j. <u>We assume that</u> $\underset{\sim}{N}\alpha$, $\underset{\sim}{N}\alpha_j$ <u>is prime to the denominator</u> N <u>of</u> a, b.

THEOREM 2.4. <u>The function</u>

$$\varphi_a(A,W)/\varphi_b(A,W)$$

<u>is a unit over</u> $\underset{\sim}{Z}$.

Finally let $\delta(W) = \Delta^{1/12}(W)$. Let

$$\delta(\alpha,W) = \delta(W)^{\underset{\sim}{N}\alpha}/\delta\left(\frac{\alpha W}{\underset{\sim}{N}\alpha}\right),$$

$$\delta(A,W) = \prod \delta(\alpha_j,W)^{n_j}.$$

Then:

<u>The product</u> $\delta(A,W)\varphi_a(A,W)$ <u>is a unit over</u> $\underset{\sim}{Z}$.

The proof is done by comparing lowest coefficients. Let $f = \sum a_n q^{n/N}$ be a power series in $q^{1/N}$, having possibly a finite number of polar terms. If g is another such power series, we write $f \sim g$ to mean that the coefficients of the lowest terms in f and g have a quotient which is a unit in a cyclotomic field. Then

$$\varphi(a_1\tau+a_2,\ \alpha,\ [\tau,1]) \sim \begin{cases} a & \text{if } a_1 \neq 0 \\ a \dfrac{\left(1 - q_1^{\langle a_2\rangle}\right)^{N\alpha}}{1 - q_1^{\langle aa_2\rangle}} & \text{if } a_1 = 0. \end{cases}$$

The theorem follows at once.

Specializing to complex multiplication, one gets Robert's units, using ideals $\underline{a}_j^{-1}$ instead of the matrices $\alpha_j/N\alpha_j$. If $L = [\omega_1, \omega_2]$, for any ideal $\underline{a}$ there is a matrix α such that $\underline{a}^{-1}L$ has the basis $\frac{\alpha W}{N\alpha}$. We can take α to be in triangular form by operating on the left with an element of $SL_2(Z)$. As a variation, if $\underline{a}_j = (\xi_j)$ is principal, then the product

$$\prod \xi_j^{-n_j}\ \underline{k}(z, \underline{a}_j^{-1}L/L)$$

is a unit in the ray class field with conductor $\underline{c}$, when L is an ideal, $z \in \underline{c}^{-1}L$, and all ideals $\underline{a}_j$ are odd, and prime to $\underline{c}$. Working with the Klein forms and the delta function separately gives a splitting of the group of units which may have arithmetic significance.

§3. REPRESENTATION OF THE CUSPS

On the modular function field F_N, the group $SL_2(\mathbb{Z}/N\mathbb{Z})$ operates in the usual way. Matrices $\sigma_d = \begin{pmatrix} 1 & 0 \\ 0 & d \end{pmatrix}$ with d prime to N operate by leaving $q^{1/N}$ fixed, and moving the coefficients of the q-expansion, $\sigma_d: \zeta_N \mapsto \zeta_N^d$. Then $GL_2(\mathbb{Z}/N\mathbb{Z})$ operates, and the isotropy group of ∞ is

$$G_\infty(N) = \left\{ \begin{pmatrix} 1 & b \\ 0 & d \end{pmatrix} \right\}.$$

There is a direct decomposition

$$GL_2(N) = C(N)G_\infty(N),$$

where $C(N)$ is described as follows. It is multiplicative over prime powers, $C(N) = \prod C(p^{n(p)})$ if $N = \prod p^{n(p)}$. The group C_p is taken to be the group of units $\mathfrak{o}_p^*$ in the unramified extension of degree 2 over $\mathbb{Z}_p$, under the regular representation. We let $C(p^n) = C_p \bmod p^n$. Then $C(p^n)$ operates simply transitively on $(\mathbb{Z}^2/p^n\mathbb{Z}^2)^* = \mathbb{Z}^2(p^n)^*$, whence on $\frac{1}{p^n}\mathbb{Z}^2/\mathbb{Z}^2$ by

$$a \mapsto a\alpha.$$

Similarly, $C(N)$ operates simply transitively on $(\frac{1}{N}\mathbb{Z}^2/\mathbb{Z}^2)^*$.

If P_∞ is the standard place at infinity of F_N, then αP_∞ for $\alpha \in C(N)/\pm 1$ gives all the other places at infinity. By this device, the cusps can be analysed with a commutative group as in complex multiplication. See [KL II].

§4. ANALYSIS OF THE REGULATOR

Let $a \in \frac{1}{N}\mathbb{Z}^2/\mathbb{Z}^2$. Let g_a be the Siegel function. Then

$$\mathrm{ord}_\infty\, g_a = \frac{1}{2}\, B_2(\langle a_1\rangle).$$

We may write the divisor of g_a in the group ring

$$(g_a) = \sum_\alpha \frac{1}{2}\, B_2(\langle T(a\alpha)\rangle)\alpha, \quad \text{with} \quad \alpha \in C(N)/\pm 1.$$

We have put $T(a_1, a_2) = a_1$. If a is primitive, so $a = (1/N, 0)\beta$ with some β, then

$$(g_a) = (g_\beta) = \sum_\alpha \frac{1}{2}\, B_2(\langle T(\beta\alpha)\rangle)\alpha.$$

Let $f(\alpha\beta)$ denote the coefficient of α in this expression. Then we have the idempotent decomposition

$$(g_\beta) = \sum_\psi \bar{\psi}(\alpha) S(\psi, f) e_\psi,$$

where ψ ranges over the character group of $C(N)/\pm 1$, e_ψ is the orthogonal idempotent of ψ, and

$$S(\psi, f) = \sum \psi(\alpha) f(\alpha).$$

Let $\psi_{\mathbb{Z}}$ be the restriction of ψ to $(\mathbb{Z}/N\mathbb{Z})^*$. Then

$$S(\psi, f) = (\text{non-zero constant}) B_{2,\psi_{\mathbb{Z}}}$$

where

$$B_{2,\psi_{\mathbb{Z}}} = N \sum_{w \in (\mathbb{Z}/N\mathbb{Z})^*} \psi_{\mathbb{Z}}(w) B_2(\langle \tfrac{w}{N}\rangle),$$

provided that the conductor of ψ is N.

The factorization of the sum $S(\psi,f)$ is carried out in [KL D], following a suggestion of Tate, using Fourier expansions of the function f. The non-zero constant consists mostly of ordinary Gauss sums, and it is known that $B_{2,\psi_Z} \neq 0$ for even characters $\psi_{\underset{\sim}{Z}}$, so the regulator of the Siegel functions is not 0.

Along the same lines, one can construct a <u>Stickelberger element</u> in the group ring, annihilating the divisor classes at infinity [KL III]. This is a 2-dimensional analogue of the situation which has arisen in the work of Leopoldt, Iwasawa, etc.

§5. DISTRIBUTION RELATIONS

Let $G = \lim G/G_n$ be a compact group, which is the projective limit of finite factor groups $X_n = G/G_n$. Let $\{\varphi_n\}$ be a family of functions, where φ_n maps X_n into some abelian group. We say that this family is consistent, or defines a <u>distribution</u> on G, if it satisfies the following condition. Let $x \in X_n$, let $\pi\colon X_{n+1} \to X_n$ be the canonical map. Then

$$\sum_{\pi y = x} \varphi_{n+1}(y) = \varphi_n(x).$$

Suppose the values of φ_n are in a ring R. We let

$$\theta_n = \sum \varphi_n(x)x \in R[G/G_n].$$

The elements θ_n are compatible under the natural homomorphisms of the group rings. We are especially interested in distributions defined by a function φ on $\mathbb{Q}^2/\mathbb{Z}^2$ satisfying the condition

$$\sum_{Ds=r} \varphi(s) = \varphi(r)$$

for every positive integer D. Such distributions are called ordinary. If the condition is satisfied only for $r \neq 0$ we say the distribution is punctured.

THEOREM 5.1. *The map* $a \longmapsto g_a$ *modulo roots of unity is an ordinary, punctured distribution.*

The relation in the case $r = 0$ has to be exceptional, since for instance g_0 is not defined. The product $\prod g_a$ for $Da = 0$, $a \neq 0$ in $\mathbb{Q}^2/\mathbb{Z}^2$ is a constant depending on D. Cf. Klein, Ramachandra [Ra] and Robert [Ro], the latter however not looking at things modularly, but through theta functions. The proofs of the relations are given from the q-expansions in [KL III].

In addition, Kubert [K 2] constructs a free basis for the units (of all level) modulo constants, showing that the distribution of Theorem 5.1 is a universal distribution (even) into abelian groups on which 2 is invertible. The combinatorics are relatively complicated, due in part to composite N, mostly to the ± 1 ambiguity. It is simple when N is a prime power.

If f is a unit, then df/f is a differential of third kind on the modular curve $X(N)$. Let us fix a determination of $\log f$ on H. For each $\gamma \in \Gamma(N)$ there is an integer $\langle f,\gamma\rangle$ in $\mathbb{Z}$ such that for all $z \in H$ we have

$$(\log f)(\gamma z) = \log f(z) + \langle f,\gamma\rangle 2\pi i.$$

Each γ thus induces a homomorphism of the group of units modulo constants into $\mathbb{Z}$, and gives rise to a homomorphic image of the Siegel distribution.

Distributions involving the higher Bernoulli numbers are discussed in [KL D], independently of the context of units in the modular function field.

§6. CHARACTERIZATION OF THE UNITS

We ask whether the construction of units by means of the Klein forms and Siegel functions gives all units in the modular function field (of all levels). The answer is yes, up to possibly 2-torsion. Let $S(N)$ be the group generated by the Siegel functions g_a with $a \in \mathbb{Q}^2$ such that $Na \in \mathbb{Z}^2$, and the non-zero constants.

THEOREM 6.1. *Let* g *be a unit.*

(i) *If* $N = p^n$ *is a prime power, and* $g^m \in S(N)$ *for some integer* $m > 0$ *then* $g \in S(N)$.

(ii) _If_ N _is composite and_ ℓ _is an odd prime, such that_ $g^\ell \in S(N)$, _then_ $g \in S(N)$. _Hence the units modulo the union of the groups_ $S(N)$ _for all_ N _form a 2-torsion group, equal to_ 1 _if_ N _is a prime power_.

The proof is in [KL IV]. It is based on a lemma of Shimura that the Fourier coefficients of a modular form have bounded denominators. Let us write

$$g = cq^\lambda g^*,$$

where c is a constant, and g^* is a power series starting with 1. The q-expansion of the Siegel functions shows that

$$g_a^* = 1 + \text{power series in } q^{1/N} \text{ with integral coeffs.}$$

Suppose

$$g^{*\ell} = \prod_a g_a^{*m(a)}$$

so that

$$g^* = \prod_a g_a^{*m(a)/\ell},$$

taking the root with the binomial series. Suppose first that N is a prime p. We look at the coefficient of $q^{1/p}$. In $g_a^{*m(a)/\ell}$ it is equal to

$$-\sum_{\nu=0}^{p-1} \frac{1}{\ell} m^*(\nu)\zeta^\nu$$

where ζ is a primitive p-th root of unity. We can choose a "good" basis for the cyclotomic integers in order to see that the coefficient has a denominator unless $g \in S(p)$. In general, when N is not prime, we look at the first non-constant coefficient and use induction.

The same method can be used to prove independence relations among the Siegel units, yielding rather easily when N is a prime power the fact that they have the proper rank as in Theorem 1.1, modulo constants. In this way one gets an independent proof that $B_{2,\chi} \neq 0$, without using the L-series. A similar method works to get the right rank when N is composite, but in a more complicated fashion.

§7. WEIERSTRASS UNITS

For $a \in \frac{1}{N}\mathbb{Z}^2/\mathbb{Z}^2$ let $\wp_a(\tau) = \wp(a_1\tau+a_2; [\tau,1])$.

Then

$$\frac{\wp_a - \wp_b}{\wp_c - \wp_d}$$

is a unit. Simple conditions can also be given on a, b c, d to determine when such an expression is a unit over Z. Note that

$$\wp_a - \wp_b = \frac{\underline{k}_{a+b}\,\underline{k}_{a-b}}{\underline{k}_a^2\,\underline{k}_b^2} .$$

Let $\underline{K}(N)$ be the group of forms expressible as products

$$\prod_a \underline{k}_a^{m(a)}$$

which are modular with respect to $\Gamma(N)$. Let $\underline{K}^+(N)$ be the subgroup of forms of even degree, i.e. such that $\sum m(a)$ is even. Let $W(N)$ be the subgroup generated by the Weierstrass forms, consisting of elements which are modular with respect to $\Gamma(N)$.

THEOREM 7.1. *We have* $W(N) = \underline{K}^+(N)$.

The proof in Kubert [K 1] is done by descent, and involves only combinatorial juggling with the quadratic relation and the parallelogram relations

$$(a+b) + (a-b) - 2(a) - 2(b).$$

§8. APPLICATIONS TO DIOPHANTINE ANALYSIS

The λ-function (generator of the modular function field of level 2) has the form of a Weierstrass unit, and so does $1-\lambda$. Put $u = \lambda$, $v = 1-\lambda$, so that $u+v = 1$. Let

$$u = x^N, \qquad v = y^N, \quad \text{so} \quad x^N + y^N = 1.$$

Fricke [F] already knew that this gives a parametrization of the Fermat curve by modular functions, which do not belong to a congruence subgroup. It gives a correspondence between $X(2N)$ and the Fermat curve $V(N)$, called *standard*.

In view of the identity

$$\frac{x_3-x_1}{x_2-x_1} + \frac{x_2-x_3}{x_2-x_1} = 1$$

we get infinitely many non-standard correspondences between the Fermat curve and modular curves. The standard one is especially good, however, because it is unramified above both $X(2N)$ and $V(N)$. From this and an old theorem of Chevalley-Weil (recalled in [KL I]), one gets:

The Fermat curve has only a finite number of rational points in any number field if and only if the modular curve $X(2N)$ *has this same property.*

In [K 3], Kubert also proves, using the Weierstrass units:

Let K *be a number field, and* ℓ *a prime* $\geqq 5$. *There is a universal bound on the torsion of elliptic curves* A *over* K *such that* ℓ *does not divide the order of the Galois group* $G(K(A_\ell)/K)$.

This follows the work of Demjanenko [D], which has not yet certifiably shown that the order of the torsion group over K is uniformly bounded, an outstanding conjecture.

The equation $u+v = 1$ has only a finite number of solutions with u, v in any finitely generated multiplicative group of complex numbers [L 3]. The Weierstrass units provide a significant example of such solutions.

The standard uniformization of the Fermat curve puts the Fermat conjecture in the light that the only rational points (over $\mathbb{Q}$!) should be at the cusps (in line with the Ogg conjecture for the modular curves belonging to congruence subgroups), cf. [KL I].

The existence of units such that $u+v = 1$ also gives a direct reduction on the modular curves for the finiteness of integral points to the situation where one can apply Gelfond's idea to use a diophantine inequality for linear combinations of logarithms of algebraic numbers, proved by Baker in general.

Let F_N be the modular function field again. *If we assume the Mordell conjecture for subfields of* F_N *properly containing* $\mathbb{Q}(j)$ *and of genus* $\geqq 2$, then we see that one gets the following irreducibility statement [KL I].

> *Let* k *be a number field containing the* N-th *roots of unity. Assume that* N *is such that every subfield of* kF_N *properly containing* $k(j)$ *has genus* $\geqq 2$ (*satisfied if* N *is prime sufficiently large*). *Then for all but a finite number of values* j_o *of* j *in* k, *we have*
>
> $$[k(\pi^{-1}(j_o)) : k] = [kF_N : k(j)],$$
>
> *where* $\pi: X(N) \longrightarrow X(1)$ *is the natural map.*

In fact, we conjecture following uniformity property as in [KL I], in connection with Serre's theorem.

<u>For all but a finite number of</u> $j_o \in k$, <u>and all but a finite number of primes</u> p, <u>the Galois group of the p-primary torsion of an elliptic curve</u> A <u>over</u> k <u>with invariant</u> j_o <u>is equal to</u> $GL_2(\mathbb{Z}_p)$.

Next we consider another aspect of torsion points.

The Manin-Mumford conjecture asserts that on a curve of genus $\geqq 2$ there are only a finite number of points of finite order in the Jacobian. The cusps on the modular curves provide significant examples of such points, according to the Manin-Drinfeld theorem (equivalent to Theorem 1.1). The question can be raised whether the cusps are also of finite order on curves which are quotients of non-congruence subgroups. This is true for the standard representation of the Fermat curve, as shown by Rohrlich [Roh], who determines completely the structure of the divisor class group generated by the cusps. On the other hand, Rohrlich has observed that the answer is negative in general. The argument goes as follows.

In [L 3], Lang reduces the Manin-Mumford conjecture to a Galois property of the field of torsion points on the Jacobian, namely that the index of the subgroup of the Galois group of the N-th torsion points over the given number field generated by the homotheties (that is, inducing multiplication by an integer prime to N on the N-th torsion points) should be bounded in $(\mathbb{Z}/N\mathbb{Z})^*$. Recently Shimura has informed us that this property can be proved in the case of complex multiplication, and therefore:

<u>The Manin-Mumford conjecture is true in the case of complex multiplication</u>.

In particular, it is true for the Fermat curve, which has complex multiplication.

By choosing infinitely many suitable non-standard correspondences of the Fermat curve with modular curves, i.e. representations as quotient of the upper half plane by non-congruence subgroups associated with units satisfying $u+v = 1$, Rohrlich shows that one would get infinitely many points on the curve of finite order in the Jacobian if the Manin-Drinfeld theorem were true in the non-congruence case, a contradiction.

APPENDIX

Because of its fundamental interest, we shall carry out here the analysis of the Weierstrass eta function $\eta(z,L)$ in detail. We recall first some facts about Eisenstein series.

By Kronecker's first limit formula or otherwise, we know that

$$E(\tau,s) = \sum \frac{y^s}{|m\tau+n|^{2s}} = \frac{\pi}{s-1} + O(1).$$

Let $L = [\omega_1,\omega_2]$, and let $\tau = \omega_1/\omega_2 \in H$. Let

$$E(L,s) = \sum_{\omega\neq 0} \frac{1}{|\omega|^{2s}}.$$

Then in a neighborhood of $s = 1$ we have

$$E(L,s) \sim \frac{\pi}{\underset{\sim}{N}L} \frac{1}{s-1},$$

where

$$\underset{\sim}{N}L = \frac{1}{2i}(\omega_1\bar{\omega}_2 - \bar{\omega}_1\omega_2)$$

is the area of the fundamental domain. The residue of $E(L,s)$ at $s = 1$ is therefore $\pi/\underset{\sim}{N}L$.

THEOREM A1. (i) The function $\sum \frac{1}{\omega^2|\omega|^{2s}}$ is holomorphic at $s = 0$ (it is even an entire function).

(ii) The function defined for $z \notin L$ by

$$G(z,L,s) = \sum_{\omega \neq 0} \frac{\overline{z+\omega}}{|z+\omega|^{2s}} \quad \text{for } \operatorname{Re} s > 3/2$$

has an analytic continuation to $\operatorname{Re} s > 1/2$, and is holomorphic at $s = 1$.

Proof. These are essentially well known, and can be proved by standard techniques using Poisson's summation formula. Cf. Siegel [Si], Theorem 3, p. 69.

In particular, we can define $s_2(L)$ as the value of the function in (i), at $s = 0$.

Following Birch-Swinnerton Dyer [B-SwD] we define

$$\zeta_s(z,L) = \frac{\bar{z}}{|z|^{2s}} + \sum_{\omega \neq 0} \left\{ \frac{\overline{z+\omega}}{|z+\omega|^{2s}} - \frac{\bar{\omega}}{|\omega|^{2s}} \left(1 - \frac{sz}{\omega} + \frac{\bar{z}}{\bar{\omega}}(1-s)\right) \right\}$$

$$= \frac{\bar{z}}{|z|^{2s}} + \sum_{\omega \neq 0} \left\{ \frac{\overline{z+\omega}}{|z+\omega|^{2s}} - \frac{\bar{\omega}}{|\omega|^{2s}} + \frac{sz}{|\omega|^{2s-2}\,\omega^2} + \frac{\bar{z}}{|\omega|^{2s}}(s-1) \right\}$$

The series converges absolutely for $\operatorname{Re} s > 1/2$ by the usual argument, and we have

$$\lim_{s \to 1} \zeta_s(z,L) = \zeta(z,L)$$

taking the limit for s real > 1. For $s > 3/2$ we can rearrange the terms in $\{\ \}$. Combining ω with $-\omega$ shows

$$\sum_{\omega \neq 0} \frac{\bar{\omega}}{|\omega|^{2s}} = 0.$$

Since $E(L,s)$ has only a simple pole at $s = 1$, it follows that

$$\sum \frac{\bar{z}}{|\omega|^{2s}} (s-1) \sim \bar{z} \frac{\pi}{N\underset{\sim}{L}} \quad \text{for } s \to 1.$$

Therefore the function

$$\sum_{\omega \neq 0} \left\{ \frac{\overline{z+\omega}}{|z+\omega|^{2s}} + \frac{sz}{|\omega|^{2s-2}\,\omega^2} \right\}$$

is holomorphic at $s = 1$. From the definition of $s_2(L)$ we then find:

$$\boxed{\zeta(z,L) = \frac{1}{z} + G(z,L,1) + s_2(L)z + \frac{\pi}{N\underset{\sim}{L}} \bar{z}.}$$

THEOREM A2. $\eta(z,L) = s_2(L)z + \frac{\pi}{\underset{\sim}{N}L}\bar{z}$.

Proof. The expression

$$\frac{1}{z} + G(z,L,1) = \frac{\bar{z}}{|z|^{2s}} + G(z,L,s) \quad \text{at} \quad s = 1$$

is periodic in L. Hence

$$\zeta(z+\omega,L) - \zeta(z,L) = s_2(L)\omega + \frac{\pi}{\underset{\sim}{N}L}\bar{\omega}.$$

$$= \eta(\omega,L).$$

The theorem follows by $\underset{\sim}{R}$-linearity. It may be viewed as a generalization of the Legendre relation, which is seen to result from the above by putting $z = \omega_1$ and $z = \omega_2$. The relation is known in the case of complex multiplication cf. Damerell, Acta Arith. XVII (1970) pp. 294 and 299, but we could find no reference for it in general.

Bibliography

[B-SwD] B. BIRCH and H. SWINNERTON-DYER, Notes on elliptic curves II, J. reine angew. Math. 218 (1965) pp. 79-108

[De] V. A. DEMJANENKO, On the uniform boundedness of the torsion of elliptic curves over algebraic number fields, Math. USSR Izvestija Vol. 6 (1972) No. 3 pp. 477-490

[Dr] V. G. DRINFELD, Two theorems on modular curves, Functional analysis and its applications, Vol. 7 No. 2, AMS translation from the Russian, April-June 1973, pp. 155-156

[F] R. FRICKE, Über die Substitutionsgruppen, welche zu den aus dem Legendre'schen Integralmodul $k^2(\omega)$ gezogenen Wurzeln gehören, Math. Ann. 28 (1887) pp. 99-118

[K 1] D. KUBERT, Quadratic relations for generators of units in the modular function field, Math. Ann. 225 (1977) pp. 1-20

[K 2] __________, A system of free generators for the universal even ordinary $\mathbf{Z}_{(2)}$ distribution on $\mathbf{Q}^{2k}/\mathbf{Z}^{2k}$, Math. Ann. 224 (1976) pp. 21-31

[K 3] __________, Universal bounds on the torsion of elliptic curves, J. London Math. Soc. to appear

[KL] D. KUBERT and S. LANG, Units in the modular function field, Math. Ann.:
I, 1975, pp. 67-96
II, 1975, pp. 175-189
III, 1975, pp. 273-285
IV, 1977, pp. 223-242

[KL D] ____________, Distributions on toroidal groups, Math. Z. 148 (1976) pp. 33-51

[L 1] S. LANG, Elliptic Functions, Addison Wesley, 1973

[L 2] _______, Division points on curves, Ann. Mat. pura et appl. IV, Tomo LXX (1965) pp. 229-234

[L 3] _______, Integral points on curves, Pub. IHES No. 6 (1960) pp. 27-43

[Ma] J. MANIN, Parabolic points and zeta functions of modular curves, Izv. Akad. Nauk SSSR, Ser. Mat. Tom 36 (1972) No. 1, AMS translation pp. 19-64

[Ner] A. NÉRON, Quasi-fonctions et hauteurs sur les variétés abéliennes, Ann. Math. 82 (1965) pp. 249-331

[New] M. NEWMAN, Construction and application of a class of modular functions, Proc. London Math. Soc. (3) (1957) pp. 334-350

[O] A. OGG, Rational points on certain elliptic modular curves, AMS conference St. Louis, 1972, pp. 221-231

[Ra] K. RAMACHANDRA, Some applications of Kronecker's limit formula, Ann. Math. 80 (1964) pp. 104-148

[Rob 1] G. ROBERT, Unités elliptiques, Bull. Soc. Math. France, Mémoire No. 36 (1973)

[Rob 2] _________, Nombres de Hurwitz et unités elliptiques, to appear

[Roh] D. ROHRLICH, Modular functions and the Fermat curve, to appear

[Si] C. L. SIEGEL, Lectures on advanced analytic number theory, Tate Institute Notes, 1961, 1965

D. Kubert
Mathematics Department
Cornell University
Ithaca, N.Y. 14850

S. Lang
Departement of Mathematics
Yale University
New Haven, Conn. 06520

CLASS FIELDS AND MODULAR FORMS OF WEIGHT ONE

by H.M. Stark

1. INTRODUCTION.

Let $f(z)$ be a normalized newform on $\Gamma_0(N)$ with a character. According to a theorem of Deligne and Serre [1] there is a normal extension K of $\mathbf{Q}$ and an irreducible two dimensional representation ρ of $\mathrm{Gal}(K/\mathbf{Q})$ such that the Artin L-series $L(s,\rho,K/\mathbf{Q})$ corresponds to $f(z)$ via a Mellin transform. Here we put forward a conjecture (see Section 3) which should aid materially in explicitly determining K from $f(z)$. Our conjecture can be proved when K is an abelian extension of a complex quadratic field and it has been numerically verified in some other instances. In the next section, we summarize the theory of Kronecker's limit formulae and complex multiplication. These results will be applied to L-series for complex quadratic fields and related via the Mellin transform to certain modular forms of weight one.

2. THE KRONECKER LIMIT FORMULAE AND COMPLEX MULTIPLICATION.

The Kronecker limit formulae are usually presented at $s = 1$. We present them here from the point of view of $s = 0$ where they would seem to be more easily applicable. Let z be in H (the upper half plane), u and v be real numbers not both integral and let $\gamma = uz + v$. We define

$$\phi(u,v,z) = e^{\pi i u\gamma}\,\frac{\theta_1(\gamma,z)}{\eta(z)}$$

where $\eta(z)$ is the Dedekind eta function and

$$\theta_1(\gamma,z) = \sum_n \exp[\pi i z(n+\tfrac{1}{2})^2 + 2\pi i(n+\tfrac{1}{2})(\gamma-\tfrac{1}{2})].$$

Let μ and ν be complex numbers with μ/ν in H and set

$$F(s) = \sum_{m,n \neq 0,0} |m\mu+n\nu|^{-2s} - 2\zeta(2s)$$

and

$$G(s) = \sum_{m,n} |m\mu+n\nu+\alpha|^{-2s}$$

where

$$\alpha = (u,v)\binom{\mu}{\nu}$$

with u and v real and not both integral. We have

$$F(0) = G(0) = 0.$$

Kronecker's first limit formula at $s = 0$ is

$$F'(0) = -4\log|\eta(\mu/\nu)|+2\log|\nu|$$

and the second limit formula is

$$G'(0) = -2\log|\phi(u,v,\mu/\nu)|.$$

Suppose $N > 1$ is an integer such that Nu and Nv are integers. Then $\phi(u,v,z)$ is a modular function of level $12N^2$ and

$$F(u,v,z) = \phi(u,v,z)^{12N}$$

is a modular function of level N [at one or two points in what follows it will be convenient to write $F_{u,v}(z)$ instead of $F(u,v,z)$]. It will be useful to recall some basic facts about such functions. Let F_N denote the field of modular functions of level N whose q-expansions at every cusp have coefficients in $\mathbf{Q}(e^{2\pi i/N})$. When Nu and Nv are integers, $\phi(u,v,z)$ is in F_{12N^2} and $F(u,v,z)$ is in F_N.

The field F_N is a normal extension of F_1 and we recall [2] that the Galois group is just $GL_2(\mathbf{Z}/N\mathbf{Z})/\pm I$. If g is in F_N and B is a 2×2 matrix of integers whose determinant is relatively prime to N, we denote by $g \circ B$ the image of g under the action of the element of $GL_2(\mathbf{Z}/N\mathbf{Z})/\pm I$ represented by B. There are two basic rules for calculating $g \circ B$. One rule is that if B is in Γ, then

$$(g \circ B)(z) = g(Bz).$$

The other rule is that if $B = \begin{pmatrix} 1 & 0 \\ 0 & p \end{pmatrix}$ where p is prime and

$$g(z) = \sum \alpha_n q^n$$

then

$$(g \circ B)(z) = \sum \alpha_n^{\sigma(p)} q^n$$

where $\sigma(p)$ is the Frobenius automorphism of $\mathbf{Q}(e^{2\pi i/N})$, associated with p :

$$\alpha^p \equiv \alpha^{\sigma(p)} (\text{mod } p).$$

There are functions in F_N which are indexed in such a way that the action by any B is easily given. The functions $F_{u,v}$ are among these. If B is a matrix of integers with determinant relatively prime to N and Nu and Nv are integers, then

$$F_{u,v} \circ B = F_{(u,v)B}.$$

The theory of complex multiplication applies to the functions of F_N and for this purpose, we need the reciprocity law. Over the years, many authors have produced versions of the reciprocity law. The best version to date is that of Shimura [4,2]. However, a somewhat less general, but still eminently usable, version which is easily proved along classical lines may be of interest. Let $d < 0$ be the discriminant of $k = \mathbf{Q}(\sqrt{d})$. If $\mathfrak{f}$ is an integral ideal of k, we denote by $K(\mathfrak{f})$ the ray class field of $k(\text{mod } \mathfrak{f})$. If $\mathfrak{p}$ is a prime ideal in k relatively prime to $\mathfrak{f}$, we denote by $\sigma(\mathfrak{p})$ the Frobenius automorphism of $K(\mathfrak{f})/k$ corresponding to $\mathfrak{p}$.

THEOREM 1. [7]. Let $g(z)$ be in F_N and suppose that $(p) = \mathfrak{p}\bar{\mathfrak{p}}$ in k where p is a rational prime such that $(p,dN) = 1$. Suppose $\mathfrak{a} = [\mu,\nu]$ is a fractional ideal of k with $\theta = \mu/\nu$ in H and let $B\binom{\mu}{\nu}$ be a basis for $\bar{\mathfrak{p}}\mathfrak{a}$. Then $g(\theta)$ is in $K(N)$ and

$$(1) \qquad g(\theta)^{\sigma(\mathfrak{p})} = [g \circ (pB^{-1})](B\theta).$$

If in addition g is analytic in the interior of H and has algebraic integer coefficients in its q-expansion at every cusp then $g(\theta)$ is an algebraic integer and we may replace (1) by

$$g(\theta)^p \equiv [g \circ (pB^{-1})](B\theta) \qquad (\text{mod } \mathfrak{p}).$$

We now apply these results to L-series at $s = 0$. Let $\mathfrak{f} \neq (1)$ be an integral ideal in $k = \mathbf{Q}(\sqrt{d})$ where $d < 0$ is the discriminant of k. If χ is a ray class character of k (mod $\mathfrak{f}$) (not necessarily primitive), we may write

$$L(s,\chi) = \sum_{C} \chi(C)z(s,C)$$

where C runs through the ray classes (mod $\mathfrak{f}$) and

$$z(s,C) = \sum_{\mathfrak{a} \in C} N(\mathfrak{a})^{-s}.$$

Let f be the minimal positive rational integer divisible by $\mathfrak{f}$ and let $w(\mathfrak{f})$ be the number of roots of unity in k which are congruent to 1 (mod $\mathfrak{f}$). Let $\mathfrak{b}$ be an integral ideal relatively prime to $\mathfrak{f}$ such that $\mathfrak{a}\mathfrak{b}$ is principal for $\mathfrak{a}$ in C. Let $\mathfrak{a}_0$ be a fixed ideal in C and $\mathfrak{a}_0\mathfrak{b} = (\alpha)$.

We have

$$w(\mathfrak{f})z(s,C) = N(\mathfrak{b})^s \sum_{\substack{\beta \in \mathfrak{b} \\ \beta \equiv \alpha \ (\text{mod } \mathfrak{f})}} N(\beta)^{-s} = N(\mathfrak{b})^s \sum_{\delta \in \mathfrak{b}\mathfrak{f}} |\delta+\alpha|^{-2s}.$$

Suppose $\mathfrak{b}\mathfrak{f} = [\mu,\nu]$ where $\theta = \mu/\nu$ is in H and $\alpha = (u,v)\binom{\mu}{\nu}$ where u and v are rational and indeed fu and fv are integral. Then by

Kronecker's second limit formula,

$$z'(0,C) = -\frac{1}{6fw(\mathfrak{f})}\log|F(u,v,\theta)|.$$

We set $E(C) = F(u,v,\theta)$. One can show that $E(C)$ is independent of the choice of α,μ,ν or even $\mathfrak{a}_0$ and $\mathfrak{b}$ and we have

(2) $$L'(0,\chi) = -\frac{1}{6fw(\mathfrak{f})}\sum_C \chi(C)\log|E(C)|.$$

By Theorem 1, the numbers $E(C)$ are in $K(f)$. But in fact they are in $K(\mathfrak{f})$. To see this, let $\mathfrak{p}$ be a first degree prime ideal in k of norm p such that $(p,df) = 1$. Let

$$\overline{\mathfrak{p}}\mathfrak{b}\mathfrak{f} = [\rho,\tau]$$

and B be defined by

$$\binom{\rho}{\tau} = B\binom{\mu}{\nu}.$$

By Theorem 1,

$$F(u,v,\theta)^p \equiv [F_{u,v}\circ(pB^{-1})](B\theta) \pmod{\mathfrak{p}}$$

$$\equiv F((u,v)pB^{-1},B\theta) \pmod{\mathfrak{p}}.$$

But the ideal $(\overline{\mathfrak{p}}\mathfrak{b})\mathfrak{f} = [\rho,\tau]$ has $\rho/\tau = B\theta$ and

$$[(u,v)pB^{-1}]\binom{\rho}{\tau} = p(u,v)\binom{\mu}{\nu} = p\alpha$$

where

$$(p\alpha) = (\mathfrak{p}\mathfrak{a}_0)(\overline{\mathfrak{p}}\mathfrak{b}).$$

Hence

$$E(C)^p \equiv E(C\mathfrak{p}) \pmod{\mathfrak{p}}.$$

Thus the $E(C)$ are all conjugate algebraic integers in $K(\mathfrak{f})$.

Let $E = E(C_1)$ where C_1 is the principal ray class (mod $\mathfrak{f}$). It is easily seen that E is a unit if and only if $\mathfrak{f}$ has two distinct prime factors. Indeed with $\chi = \chi_1$, the principal (imprimitive) character (mod $\mathfrak{f}$), we have from (2),

$$\log(N_{K(\mathfrak{f})/\mathbf{Q}}E) = -12fw(\mathfrak{f})L'(0,\chi_1)$$

where the derivative of

$$L(s,\chi_1) = \zeta_k(s) \prod_{p|\mathfrak{f}} (1 - N(p)^{-s})$$

is 0 at s = 0 if and only if $\mathfrak{f}$ has two distinct prime factors.

When the numbers E(C) are units, the question arises as to the index of the subgroup of the unit group of $K(\mathfrak{f})$ that the E(C) and certain other units generate. See [3] for example. The factor of 6f in the denominator of (2) makes this index annoyingly high. The explicit reciprocity law allows us to improve this state of affairs considerably.

LEMMA 1. *Let ℓ be a first degree prime ideal in* k *of norm l with* $(l,12fd) = 1$ *and suppose ℓ is in the ray class* C_2 (mod $\mathfrak{f}$). *Then* $E(C_2)/E^l$ *is the* $(12f)^{th}$ *power of a number in* $K(\mathfrak{f})$.

PROOF. Up to a root of unity, we found that

$$E(C)^{1/(12f)} = \phi(u,v,\theta)$$

and $\phi(u,v,z)$ is a modular function in F_{12f^2}. Hence $E(C)^{1/(12f)}$ is in $K(12f^2)$, a field which incidentally contains the $(12f)^{th}$ roots of unity. Let δ denote one of the $(12f)^{th}$ roots of E and set $\delta(\ell) = \delta^{\sigma(\ell)}$ where $\sigma(\ell)$ denotes the Frobenius automorphism in $K(12f^2)$ corresponding to ℓ. In particular $\delta(\ell)^{12f} = E(C_2)$. Let p be a first degree prime ideal in k of norm p such that p is in C_1 and $(p,12fd) = 1$. Since δ^{12f} is preserved by $\sigma(p)$, we see that $\delta^{\sigma(p)} = \omega\delta$ where $\omega^{12f} = 1$. But now,

$$\delta(\ell)^{\sigma(p)} = \delta^{\sigma(\ell)\sigma(p)} = \delta^{\sigma(p)\sigma(\ell)} = (\omega\delta)^{\sigma(\ell)} = \omega^l\delta(\ell)$$

and so $\sigma(p)$ preserves $\delta(\ell)/\delta^l$ and the lemma follows.

COROLLARY 1. *Suppose* $K(\mathfrak{f})$ *contains exactly* W *roots of unity. Then* E^W *is a* $(12f)^{th}$ *power of an integer in* $K(\mathfrak{f})$.

PROOF. We use the criterion that a necessary and sufficient condition for $K(\mathfrak{f})$ to contain the mth roots of unity is that

$$N(\ell) \equiv 1 \pmod{m}$$

for all first degree prime ideals ℓ in C_1 with $(N(\ell),m) = 1$. Thus W is the greatest common divisor of the set of numbers $N(\ell) - 1$ as ℓ runs through those first degree prime ideals of k in C_1 with $(N(\ell),W) = 1$. The corollary now follows from Lemma 1 and the Euclidean algorithm.

THEOREM 2. Suppose $K(\mathfrak{f})$ contains exactly W roots of unity. Then there are conjugate algebraic integers $\varepsilon(C)$ in $K(\mathfrak{f})$ such that

$$L'(0,\chi) = -\frac{1}{Ww(\mathfrak{f})}\sum_{C} \chi(C)\log|\varepsilon(C)|^2$$

(Note that $w(\mathfrak{f})$ is usually 1.)

The proof is immediate.

Similar results are possible for fields intermediate to $K(\mathfrak{f})$ and k. For details on this and other questions related to this section, see [7].

3. FORMS OF WEIGHT ONE.

Again, let $\mathfrak{f} \neq (1)$ be an integral ideal in $\mathbf{Q}(\sqrt{d})$ where $d < 0$ is the discriminant of $\mathbf{Q}(\sqrt{d})$ and let χ be a not necessarily primitive character (mod $\mathfrak{f}$). Define $g_\chi(z)$ by the Mellin transform,

$$(2\pi)^{-s}\Gamma(s)L(s,\chi) = \int_0^\infty y^{s-1} g_\chi(iy)dy.$$

Thus $g_\chi(z)$ is a modular form of weight one on $\Gamma_1(N)$ with $N = |d|N(\mathfrak{f})$. [This is true even when χ is not primitive. Eliminating the p

factor of $L(s,\chi)$ is accomplished by multiplying $L(s,\chi)$ by $(1-\chi(p)N(p)^{-s})$ and this corresponds to replacing $g_\chi(z)$ by $g_\chi(z)-\chi(p)g_\chi(N(p)z)$ which raises the level by a factor of $N(p)$ at worst.] In particular,

$$L'(0,\chi) = \int_0^\infty g_\chi(iy)\frac{dy}{y}.$$

When we compare this with the results of the last section, we are led to the following conjecture.

CONJECTURE 1. *Let* $f(z)$ *be a modular form of weight one on* $\Gamma_1(N)$ *which is* 0 *at* $i\infty$. *Then*

$$\int_0^\infty f(iy)\frac{dy}{y} = \sum_{j=1}^{n} \rho_j \log \varepsilon_j$$

where the ε_j *are algebraic integers and the* ρ_j *lie in the field generated over* $\mathbf{Q}$ *by adjoining the coefficients of the* q-*expansion of* $f(z)$ *at* $i\infty$.

Except for the problem that the field given by adjoining the values of $\chi(C)$ to $\mathbf{Q}$ may be slightly bigger than the field given by adjoining the coefficients of g_χ to $\mathbf{Q}$, we proved this conjecture for g_χ in Theorem 2. Of course $\mathfrak{f} = (1)$ may be allowed also provided we exclude $\chi = \chi_1$, the principal character (the corresponding form is not 0 at $i\infty$). As an example, let χ be either one of the two cubic ideal class characters of $\mathbf{Q}(\sqrt{-23})$ so that

$$g_\chi(z) = \eta(z)\eta(23z).$$

Here the q-expansion of g_χ has rational coefficients (the values of χ generate $\mathbf{Q}(\sqrt{-3})$) and

$$\int_0^\infty g_\chi(iy)\frac{dy}{y} = L'(0,\chi) = \log \varepsilon$$

where ε is the real root of

$$x^3 - x - 1 = 0.$$

The principal interest in Conjecture 1 lies in determining the field containing the ε_j. In particular, it should ultimately lead to a constructive version of the Deligne-Serre theorem. To make this clearer, let f(z) be a normalized newform on $\Gamma_1(N)$. Thus f(z) is actually defined on $\Gamma_0(N)$ with a character and according to the theorem of Deligne and Serre there is a normal extension K of $\mathbf{Q}$ and an irreducible two dimensional representation ρ of $Gal(K/\mathbf{Q})$ such that the Dirichlet series corresponding to f(z) is just the Artin L-series $L(s,\chi,K/\mathbf{Q})$ where χ is the character of $Gal(K/\mathbf{Q})$ corresponding to ρ.

Let $\alpha_1,\ldots,\alpha_n$ be an integral basis of the field giving by adjoining the values of χ to $\mathbf{Q}$ and set

$$f(z) = \sum_{j=1}^{n} \alpha_j g_j(z)$$

so that the $g_j(z)$ are also cusp forms on $\Gamma_1(N)$ with rational coefficients. A refined version of Conjecture 1 is

CONJECTURE 2. *There is a rational integer* m *such that*

$$\int_0^\infty g_j(iy)\frac{dy}{y} = \frac{1}{m}\log \varepsilon_j$$

where ε_j *is an integer in the maximal real subfield of* K.

All available evidence says that the number m is closely related to the number of roots of unity in K. Note that the numbers $|\varepsilon(C)|^2$ of the previous section lie in the minimal normal extension of $\mathbf{Q}$ containing $K(\zeta)$. As a numerical verification of Conjecture 2, let N = 145. We take f(z) to be the modular form corresponding to the Dirichlet series $L(s,\chi)$ on $\mathbf{Q}(\sqrt{5})$ where χ is the ray class character of order 4 (mod $\mathfrak{f}\mathfrak{p}_\infty$), $\mathfrak{f} = \left(\frac{11-\sqrt{5}}{2}\right)$, given by $\chi((2)) = i$. We have

$$f(z) = g(z) + ih(z)$$

where g and h are on $\Gamma_1(145)$ with rational coefficients. We find for example,

$$\int_0^\infty g(iy)\frac{dy}{y} = 1.656074962913147\cdots \quad . \tag{3}$$

Let

$$\varepsilon = \frac{(3+2\sqrt{5}) + \sqrt{7+2\sqrt{5}} + \sqrt{(20+14\sqrt{5}) + (6+4\sqrt{5})\sqrt{7+2\sqrt{5}}}}{4}$$

so that ε is a unit in the ray class field (mod $\mathfrak{f}\mathfrak{p}_\infty$) of $\mathbf{Q}(\sqrt{5})$. We have

$$\log \varepsilon = 1.656074962913158\cdots \quad . \tag{4}$$

The discrepancies between (3) and (4) at the end are presumably due to round off errors on the computer which did all calculations to 16 decimal places. This example was discussed more fully in my Durham talk [5] from another viewpoint. In particular, it was shown how the numerical values of the integral in (3) and the corresponding integral for h can be used to find ε and why this ε is more reasonable than any of the other infinitely many units close to ε. There is considerable numerical evidence [5,6] for Conjecture 2 when the modular form f comes from a real quadratic field. A meaningful numerical verification for $N = 133$ would be of some interest.

REFERENCES

[1] DELIGNE P. and SERRE J-P., Formes modulaires de poids 1, Ann. Scient. Éc. Norm. Sup. (4), 7 (1974), 507-530.

[2] LANG S., Elliptic Functions, Addison Wesley, Reading, Mass. 1973.

[3] ROBERT G., Unités elliptiques et formules pour le nombre de classes des extensions abéliennes d'un corps quadratique imaginaire, Bull. Soc. Math. de France, Mémoire no. 36, 1973.

[4] SHIMURA G., Introduction to the Arithmetic Theory of Automorphic Functions, Iwanami Shoten and Princeton University Press, 1971.

[5] STARK H.M., Class fields for real quadratic fields and L-series at 1, To appear in Proc. of the 1975 Durham conference.

[6] STARK H.M., L-series at s = 1, III. Totally real fields and Hilbert's twelfth problem, Advances in Math. 22 (1976), 64-84.

[7] STARK H.M., L-series at s = 1, IV, To be published.

H. Stark
Department of Mathematics
Massachusetts Institute
of Technology
Cambridge, Mass. 02139

An Icosahedral Modular Form of Weight One

Joe Buhler

It is conjectured that there is a one-to-one correspondence between odd two-dimensional galois representations $\rho: \mathrm{Gal}(\bar{\mathbb{Q}}/\mathbb{Q}) \longrightarrow GL_2(\mathbb{C})$ of conductor N and determinant ϵ, and newforms f on $\Gamma_0(N)$ of weight one and character ϵ. The L-series $L(s,\rho)$ associated (by Artin) to a galois representation should coincide with the L-series $L_f(s)$ associated (by Hecke) to the corresponding modular form f. For a precise discussion of the situation (and a description of the notation) the reader should consult [2].

A two-dimensional galois representation over $\mathbb{Q}$ determines a finite galois extension L of $\mathbb{Q}$ together with a faithful representation $\mathrm{Gal}(L/\mathbb{Q}) \longrightarrow GL_2(\mathbb{C})$ of the galois group of $L/\mathbb{Q}$. If K is the fixed field of the kernel of the composition

$$\mathrm{Gal}(L/\mathbb{Q}) \longrightarrow GL_2(\mathbb{C}) \longrightarrow PGL_2(\mathbb{C}) = GL_2(\mathbb{C})/\mathbb{C}^*$$

then we get a diagram as follows:

$$1) \qquad \begin{array}{ccccccccc} 0 & \longrightarrow & \mathrm{Gal}(L/K) & \longrightarrow & \mathrm{Gal}(L/\mathbb{Q}) & \longrightarrow & \mathrm{Gal}(K/\mathbb{Q}) & \longrightarrow & 0 \\ & & \downarrow & & \downarrow & & \downarrow & & \\ 0 & \longrightarrow & \mathbb{C}^* & \longrightarrow & GL_2(\mathbb{C}) & \longrightarrow & PGL_2(\mathbb{C}) & \longrightarrow & 0 \end{array}$$

The finite subgroups of $PGL_2(\mathbb{C})$ have been classified since the days of Klein; such a group is either cyclic, dihedral or the group of symmetries of one of the Platonic solids: the tetrahedral group (isomorphic to A_4), the octahedral group (isomorphic to S_4), or the icosahedral group (isomorphic to A_5).

If f is a cuspidal newform of level N and character ϵ (so that, in particular, f is an eigenform of all of the Hecke operators T_p for $p \nmid N$) then a theorem of Deligne and Serre (see [2]) implies the existence of an irreducible odd galois representation $\rho:\mathrm{Gal}(\overline{\mathbb{Q}}/\mathbb{Q}) \longrightarrow GL_2(\mathbb{C})$ such that $L(s,\rho) = L_f(s)$. If ρ is an irreducible monomial representation (= induced from a one-dimensional representation) then the corresponding projective group is dihedral; in this case call the modular form f a dihedral form. Similarly call the form f tetrahedral (resp. octahedral, icosahedral) if the representation ρ is tetrahedral (resp. octahedral, icosahedral).

Results of Weil and Langlands, together with Langlands' results on the constants in the functional equation for $L(s,\rho)$, imply that if Artin's conjecture is true for a galois representation ρ and all of its "twists" $\rho\otimes\psi$ by one-dimensional representations ψ, then there is a modular form f as above such that $L_f(s) = L(s,\rho)$. The validity of Artin's conjecture for reducible representations (= those representations whose associated projective groups are cyclic) is classical; similarly for monomial representations. Recently Langlands (see [5]) has shown that Artin's conjecture is true for tetrahedral representations and for "half" of the octahedral representations. Thus in all of these cases the one-to-one correspondence alluded to above has been established.

The ideas used in [5] pivot around the fact that the galois groups associated to tetrahedral and octahedral representations are solvable (for an idea of the techniques involved see [3]). The object of this note is to discuss a proof of Artin's conjecture for one case of an icosahedral (and thus non-solvable) representation. More specifically, we 1) outline the steps of the proof of the existence of an icosahedral form, and 2) state a general result on conductors of local galois representations obtained in the course of finding the icosahedral form. The details can be found in [1].

Theorem 1: *There is an icosahedral form of level* 800.

Corollary: *There is an icosahedral representation* $\rho:\mathrm{Gal}(\overline{\mathbb{Q}}/\mathbb{Q}) \longrightarrow GL_2(\mathbb{C})$ *of conductor* 800 *such that* $L(s,\rho)$ *is an entire function.*

Remark: The corollary follows immediately from the theorem by the result of Deligne and Serre together with the fact that $L_f(s)$ is holomorphic for every cuspidal newform f.

The first step of the proof of the theorem is the construction of an icosahedral galois representation. An icosahedral representation determines an A_5 extension K of $\mathbf{Q}$ as in diagram 1) above. It is natural to describe K by giving a quintic polynomial with integral coefficients such that K is the splitting field of this polynomial.

Given such a quintic polynomial, the next step is to calculate the minimal conductor of a representation $\rho:\mathrm{Gal}(L/\mathbf{Q}) \longrightarrow GL_2(\mathbf{C})$ that "lifts" the projective representation $\mathrm{Gal}(K/\mathbf{Q}) \longrightarrow PGL_2(\mathbf{C})$. (In fact that are two non-isomorphic two-dimensional projective representations of A_5; the conductor of a lifting is independent of which projective representation is chosen.)

According to some results of Tate the minimal conductor problem reduces to a collection of local problems (for a statement of these results and a general discussion of this "lifting" problem see [7]). Thus we are given a galois extension K_p of $\mathbf{Q}_p$ and a projective representation $\bar{\rho}:\mathrm{Gal}(K_p/\mathbf{Q}_p) \longrightarrow PGL_2(\mathbf{C})$, and we want to find a linear representation of minimal conductor that lifts $\bar{\rho}$. The most difficult case is when $\bar{\rho}$ is primitive (which means that any linear representation that lifts $\bar{\rho}$ is not monomial).

The ideas necessary to find the local minimal conductor in this case generalize easily to the case of primitive representations of prime degree over an arbitrary local field. The degree of such a representation must be equal to the residue characteristic of the local field, and the image of the wild ramification group must be an elementary abelian group whose order is the square of the residue characteristic. For a description of the images of primitive projective representations of arbitrary degree see [4].

<u>Theorem 2: Let</u> F <u>be a local field of residue characteristic</u> p. <u>Let</u>

$\overline{\rho}$:Gal(K/F) ⟶ $PGL_p(\mathbf{C})$ _be a faithful primitive projective representation of degree_ p. _Let_ e _be the tame ramification index of_ K/F, _and let_ r _be the largest integer such that_ $G_r \neq \{0\}$, _where_ G = Gal(K/F) _and_ $\{G_n\}$ _is the sequence of ramification subgroup of_ G _in the "lower numbering". Then the minimal conductor of a lifting_ ρ:Gal($\overline{F}$/F) ⟶ $GL_p(\mathbf{C})$ _of_ $\overline{\rho}$ _is_

$$p + \left(\frac{p+1}{e}\right)r.$$

In order to find candidates for icosahedral representations that might yield icosahedral forms I first used a computer to search for appropriate quintic polynomials. The machine sieved through many (about 10^7) quintics and discarded those whose galois groups were not A_5. Then the minimal conductor associated to the splitting field of the quintic was calculated using Theorem 2 together with an analysis of the reducible and monomial local representations. The smallest conductors that were found are as follows:

$800 = 2^5 5^2$	$1188 = 2^2 3^3 11$	$1687 = 7\ 241$
$837 = 3^3 31$	$1376 = 2^5 43$	$1825 = 5^2 73$
$992 = 2^5 31$	$1501 = 19\ 79$	$1948 = 2^2 487$
$1161 = 3^3 43$	$1600 = 2^6 5^2$	2083 = a prime

The quintic with "conductor" 800 was chosen for detailed consideration not only because it gave the lowest conductor but also because of the interesting local behavior of the representation at 2 and 5; e.g., the local galois extension of the quintic at p = 2 is the unique tetrahedral extension $K_2/\mathbf{Q}_2$ of $\mathbf{Q}_2$ (see [9]).

The next stage in the proof of Theorem 1 is the calculation of some of the coefficients of the L-series L(s, ρ) of a representation ρ attached to the specific quintic field. This requires a description of the extension L/K as in diagram 1). Fortunately L/K "comes from" a cyclic extension L'/E, where

E is a subfield of K fixed by a dihedral subgroup of A_5 of order 10. Thus E is a sextic extension of $\mathbf{Q}$; in fact E is the classical sextic resolvent field associated to the given A_5 polynomial. The calculation of $L(s, \rho)$ essentially reduces to some computations in a generalized ideal class group of E.

Let $f = \Sigma a_n q^n$ be the "Mellin transform" of the Dirichlet series $L(s,\rho) = \Sigma a_n n^{-s}$; in principle, the ideas above enable one to compute any desired initial segment of the power series f. Hence we can follow the technique of [8] and take a known modular form of weight one and level 800, say g, and multiply the q-expansion of g by f. If g is chosen so that its character is the inverse of the Dirichlet character $\det(\rho)$, then one would hope to have

2) $$fg = h$$

for some modular form h of weight 2 (and trivial character) on $\Gamma_0(800)$.

In order to carry out this computation explicitly it is necessary to 1) construct a basis for the (97-dimensional) space of forms of weight 2, and 2) know the first 360 coefficients of f. If we then let f' be the quotient of h by g, then f' is a modular form as in Theorem 1 except that it might not be holomorphic at the zeroes of g, and it might not be an eigenform for the Hecke operators T_p ($p \neq 2,5$). With some work we can actually show that f' is holomorphic by considering several different g_i's, finding $f'g_i = h_i$, and then showing that the g_i have no common zero. Moreover enough coefficients of f' are known to show that it is an eigenform for the Hecke operator T_3. It can be checked that there are no tetrahedral or octahedral representations of conductor dividing 800, and hence (by Deligne-Serre) no tetrahedral or octahedral forms of level 800. It is also easy to enumerate the dihedral forms of this level and character; their eigenvalues for T_3 are unequal to the eigenvalue of f' under the action of T_3. The only possibility is that there must be an icosahedral form as claimed in the theorem.

The proof shows only that the first 360 coefficients of the q-expansion of f' agree with the corresponding initial segment of the power series f obtained from $L(s, \rho)$; in particular, this does not show that the representation ρ that we started with is the same as the representation of the Corollary. Using the technique of Odlyzko-Poitou-Serre (see [6]) for bounding discriminants it is possible to show that if the Generalized Riemann Hypothesis is true (resp. if f' is an eigenform for T_{11}) then ρ is the only galois representation of conductor 800 and with the given determinant (resp. the only galois representation of conductor 800, the given determinant, and the given L-series coefficients a_3 and a_{11}).

References:

[1] J. Buhler, thesis, Harvard University, 1977.

[2] P. Deligne and J.-P. Serre, Formes modulaires de poids 1, Ann. Scient. Ec. Norm. Sup. (4), 7 (1974), p. 507.

[3] S. Gelbart, Automorphic Forms and Artin's Conjecture, these proceedings.

[4] H. Koch, Classification of the primitive representations of the Galois groups of local fields (preprint).

[5] R. P. Langlands, Base Change for GL(2), Notes for Lectures at the Institute for Advanced Study, 1975.

[6] G. Poitou, Minorations de discriminants, Sem. Bourbaki, 479, 1976.

[7] J.-P. Serre, Modular forms of weight one and galois representations, Proceedings of a conference at Durham, to appear.

[8] J. Tate, various notes on the work of Tate, Atkin, Flath, Kottwitz, Tunnell and Weisinger on the tetrahedral representation of conductor 133 (unpublished).

[9] A. Weil, Exercices Dyadiques, Invent. Math., 27 (1974) p.1.

Dr. J. Buhler
Harvard University
Dept. of Mathematics
Science Center
One Oxford Street
Cambridge, Mass. 02138

Vol. 460: O. Loos, Jordan Pairs. XVI, 218 pages. 1975.

Vol. 461: Computational Mechanics. Proceedings 1974. Edited by J. T. Oden. VII, 328 pages. 1975.

Vol. 462: P. Gérardin, Construction de Séries Discrètes p-adiques. »Sur les séries discrètes non ramifiées des groupes réductifs déployés p-adiques«. III, 180 pages. 1975.

Vol. 463: H.-H. Kuo, Gaussian Measures in Banach Spaces. VI, 224 pages. 1975.

Vol. 464: C. Rockland, Hypoellipticity and Eigenvalue Asymptotics. III, 171 pages. 1975.

Vol. 465: Séminaire de Probabilités IX. Proceedings 1973/74. Edité par P. A. Meyer. IV, 589 pages. 1975.

Vol. 466: Non-Commutative Harmonic Analysis. Proceedings 1974. Edited by J. Carmona, J. Dixmier and M. Vergne. VI, 231 pages. 1975.

Vol. 467: M. R. Essén, The Cos $\pi\lambda$ Theorem. With a paper by Christer Borell. VII, 112 pages. 1975.

Vol. 468: Dynamical Systems – Warwick 1974. Proceedings 1973/74. Edited by A. Manning. X, 405 pages. 1975.

Vol. 469: E. Binz, Continuous Convergence on C(X). IX, 140 pages. 1975.

Vol. 470: R. Bowen, Equilibrium States and the Ergodic Theory of Anosov Diffeomorphisms. III, 108 pages. 1975.

Vol. 471: R. S. Hamilton, Harmonic Maps of Manifolds with Boundary. III, 168 pages. 1975.

Vol. 472: Probability-Winter School. Proceedings 1975. Edited by Z. Ciesielski, K. Urbanik, and W. A. Woyczyński. VI, 283 pages. 1975.

Vol. 473: D. Burghelea, R. Lashof, and M. Rothenberg, Groups of Automorphisms of Manifolds. (with an appendix by E. Pedersen) VII, 156 pages. 1975.

Vol. 474: Séminaire Pierre Lelong (Analyse) Année 1973/74. Edité par P. Lelong. VI, 182 pages. 1975.

Vol. 475: Répartition Modulo 1. Actes du Colloque de Marseille-Luminy, 4 au 7 Juin 1974. Edité par G. Rauzy. V, 258 pages. 1975. 1975.

Vol. 476: Modular Functions of One Variable IV. Proceedings 1972. Edited by B. J. Birch and W. Kuyk. V, 151 pages. 1975.

Vol. 477: Optimization and Optimal Control. Proceedings 1974. Edited by R. Bulirsch, W. Oettli, and J. Stoer. VII, 294 pages. 1975.

Vol. 478: G. Schober, Univalent Functions – Selected Topics. V, 200 pages. 1975.

Vol. 479: S. D. Fisher and J. W. Jerome, Minimum Norm Extremals in Function Spaces. With Applications to Classical and Modern Analysis. VIII, 209 pages. 1975.

Vol. 480: X. M. Fernique, J. P. Conze et J. Gani, Ecole d'Eté de Probabilités de Saint-Flour IV-1974. Edité par P.-L. Hennequin. XI, 293 pages. 1975.

Vol. 481: M. de Guzmán, Differentiation of Integrals in R^n. XII, 226 pages. 1975.

Vol. 482: Fonctions de Plusieurs Variables Complexes II. Séminaire François Norguet 1974–1975. IX, 367 pages. 1975.

Vol. 483: R. D. M. Accola, Riemann Surfaces, Theta Functions, and Abelian Automorphisms Groups. III, 105 pages. 1975.

Vol. 484: Differential Topology and Geometry. Proceedings 1974. Edited by G. P. Joubert, R. P. Moussu, and R. H. Roussarie. IX, 287 pages. 1975.

Vol. 485: J. Diestel, Geometry of Banach Spaces – Selected Topics. XI, 282 pages. 1975.

Vol. 486: S. Stratila and D. Voiculescu, Representations of AF-Algebras and of the Group U (∞). IX, 169 pages. 1975.

Vol. 487: H. M. Reimann und T. Rychener, Funktionen beschränkter mittlerer Oszillation. VI, 141 Seiten. 1975.

Vol. 488: Representations of Algebras, Ottawa 1974. Proceedings 1974. Edited by V. Dlab and P. Gabriel. XII, 378 pages. 1975.

Vol. 489: J. Bair and R. Fourneau, Etude Géométrique des Espaces Vectoriels. Une Introduction. VII, 185 pages. 1975.

Vol. 490: The Geometry of Metric and Linear Spaces. Proceedings 1974. Edited by L. M. Kelly. X, 244 pages. 1975.

Vol. 491: K. A. Broughan, Invariants for Real-Generated Uniform Topological and Algebraic Categories. X, 197 pages. 1975.

Vol. 492: Infinitary Logic: In Memoriam Carol Karp. Edited by D. W. Kueker. VI, 206 pages. 1975.

Vol. 493: F. W. Kamber and P. Tondeur, Foliated Bundles and Characteristic Classes. XIII, 208 pages. 1975.

Vol. 494: A Cornea and G. Licea. Order and Potential Resolvent Families of Kernels. IV, 154 pages. 1975.

Vol. 495: A. Kerber, Representations of Permutation Groups II. V, 175 pages. 1975.

Vol. 496: L. H. Hodgkin and V. P. Snaith, Topics in K-Theory. Two Independent Contributions. III, 294 pages. 1975.

Vol. 497: Analyse Harmonique sur les Groupes de Lie. Proceedings 1973–75. Edité par P. Eymard et al. VI, 710 pages. 1975.

Vol. 498: Model Theory and Algebra. A Memorial Tribute to Abraham Robinson. Edited by D. H. Saracino and V. B. Weispfenning. X, 463 pages. 1975.

Vol. 499: Logic Conference, Kiel 1974. Proceedings. Edited by G. H. Müller, A. Oberschelp, and K. Potthoff. V, 651 pages 1975.

Vol. 500: Proof Theory Symposion, Kiel 1974. Proceedings. Edited by J. Diller and G. H. Müller. VIII, 383 pages. 1975.

Vol. 501: Spline Functions, Karlsruhe 1975. Proceedings. Edited by K. Böhmer, G. Meinardus, and W. Schempp. VI, 421 pages. 1976.

Vol. 502: János Galambos, Representations of Real Numbers by Infinite Series. VI, 146 pages. 1976.

Vol. 503: Applications of Methods of Functional Analysis to Problems in Mechanics. Proceedings 1975. Edited by P. Germain and B. Nayroles. XIX, 531 pages. 1976.

Vol. 504: S. Lang and H. F. Trotter, Frobenius Distributions in GL_2-Extensions. III, 274 pages. 1976.

Vol. 505: Advances in Complex Function Theory. Proceedings 1973/74. Edited by W. E. Kirwan and L. Zalcman. VIII, 203 pages. 1976.

Vol. 506: Numerical Analysis, Dundee 1975. Proceedings. Edited by G. A. Watson. X, 201 pages. 1976.

Vol. 507: M. C. Reed, Abstract Non-Linear Wave Equations. VI, 128 pages. 1976.

Vol. 508: E. Seneta, Regularly Varying Functions. V, 112 pages. 1976.

Vol. 509: D. E. Blair, Contact Manifolds in Riemannian Geometry. VI, 146 pages. 1976.

Vol. 510: V. Poènaru, Singularités C^∞ en Présence de Symétrie. V, 174 pages. 1976.

Vol. 511: Séminaire de Probabilités X. Proceedings 1974/75. Edité par P. A. Meyer. VI, 593 pages. 1976.

Vol. 512: Spaces of Analytic Functions, Kristiansand, Norway 1975. Proceedings. Edited by O. B. Bekken, B. K. Øksendal, and A. Stray. VIII, 204 pages. 1976.

Vol. 513: R. B. Warfield, Jr. Nilpotent Groups. VIII, 115 pages. 1976.

Vol. 514: Séminaire Bourbaki vol. 1974/75. Exposés 453 – 470. IV, 276 pages. 1976.

Vol. 515: Bäcklund Transformations. Nashville, Tennessee 1974. Proceedings. Edited by R. M. Miura. VIII, 295 pages. 1976.

Vol. 516: M. L. Silverstein, Boundary Theory for Symmetric Markov Processes. XVI, 314 pages. 1976.

Vol. 517: S. Glasner, Proximal Flows. VIII, 153 pages. 1976.

Vol. 518: Séminaire de Théorie du Potentiel, Proceedings Paris 1972–1974. Edité par F. Hirsch et G. Mokobodzki. VI, 275 pages. 1976.

Vol. 519: J. Schmets, Espaces de Fonctions Continues. XII, 150 pages. 1976.

Vol. 520: R. H. Farrell, Techniques of Multivariate Calculation. X, 337 pages. 1976.

Vol. 521: G. Cherlin, Model Theoretic Algebra – Selected Topics. IV, 234 pages. 1976.

Vol. 522: C. O. Bloom and N. D. Kazarinoff, Short Wave Radiation Problems in Inhomogeneous Media: Asymptotic Solutions. V. 104 pages. 1976.

Vol. 523: S. A. Albeverio and R. J. Høegh-Krohn, Mathematical Theory of Feynman Path Integrals. IV, 139 pages. 1976.

Vol. 524: Séminaire Pierre Lelong (Analyse) Année 1974/75. Edité par P. Lelong. V, 222 pages. 1976.

Vol. 525: Structural Stability, the Theory of Catastrophes, and Applications in the Sciences. Proceedings 1975. Edited by P. Hilton. VI, 408 pages. 1976.

Vol. 526: Probability in Banach Spaces. Proceedings 1975. Edited by A. Beck. VI, 290 pages. 1976.

Vol. 527: M. Denker, Ch. Grillenberger, and K. Sigmund, Ergodic Theory on Compact Spaces. IV, 360 pages. 1976.

Vol. 528: J. E. Humphreys, Ordinary and Modular Representations of Chevalley Groups. III, 127 pages. 1976.

Vol. 529: J. Grandell, Doubly Stochastic Poisson Processes. X, 234 pages. 1976.

Vol. 530: S. S. Gelbart, Weil's Representation and the Spectrum of the Metaplectic Group. VII, 140 pages. 1976.

Vol. 531: Y.-C. Wong, The Topology of Uniform Convergence on Order-Bounded Sets. VI, 163 pages. 1976.

Vol. 532: Théorie Ergodique. Proceedings 1973/1974. Edité par J.-P. Conze and M. S. Keane. VIII, 227 pages. 1976.

Vol. 533: F. R. Cohen, T. J. Lada, and J. P. May, The Homology of Iterated Loop Spaces. IX, 490 pages. 1976.

Vol. 534: C. Preston, Random Fields. V, 200 pages. 1976.

Vol. 535: Singularités d'Applications Differentiables. Plans-sur-Bex. 1975. Edité par O. Burlet et F. Ronga. V, 253 pages. 1976.

Vol. 536: W. M. Schmidt, Equations over Finite Fields. An Elementary Approach. IX, 267 pages. 1976.

Vol. 537: Set Theory and Hierarchy Theory. Bierutowice, Poland 1975. A Memorial Tribute to Andrzej Mostowski. Edited by W. Marek, M. Srebrny and A. Zarach. XIII, 345 pages. 1976.

Vol. 538: G. Fischer, Complex Analytic Geometry. VII, 201 pages. 1976.

Vol. 539: A. Badrikian, J. F. C. Kingman et J. Kuelbs, Ecole d'Eté de Probabilités de Saint Flour V-1975. Edité par P.-L. Hennequin. IX, 314 pages. 1976.

Vol. 540: Categorical Topology, Proceedings 1975. Edited by E. Binz and H. Herrlich. XV, 719 pages. 1976.

Vol. 541: Measure Theory, Oberwolfach 1975. Proceedings. Edited by A. Bellow and D. Kölzow. XIV, 430 pages. 1976.

Vol. 542: D. A. Edwards and H. M. Hastings, Čech and Steenrod Homotopy Theories with Applications to Geometric Topology. VII, 296 pages. 1976.

Vol. 543: Nonlinear Operators and the Calculus of Variations, Bruxelles 1975. Edited by J. P. Gossez, E. J. Lami Dozo, J. Mawhin, and L. Waelbroeck, VII, 237 pages. 1976.

Vol. 544: Robert P. Langlands, On the Functional Equations Satisfied by Eisenstein Series. VII, 337 pages. 1976.

Vol. 545: Noncommutative Ring Theory. Kent State 1975. Edited by J. H. Cozzens and F. L. Sandomierski. V, 212 pages. 1976.

Vol. 546: K. Mahler, Lectures on Transcendental Numbers. Edited and Completed by B. Diviš and W. J. Le Veque. XXI, 254 pages. 1976.

Vol. 547: A. Mukherjea and N. A. Tserpes, Measures on Topological Semigroups: Convolution Products and Random Walks. V, 197 pages. 1976.

Vol. 548: D. A. Hejhal, The Selberg Trace Formula for PSL (2,IR). Volume I. VI, 516 pages. 1976.

Vol. 549: Brauer Groups, Evanston 1975. Proceedings. Edited by D. Zelinsky. V, 187 pages. 1976.

Vol. 550: Proceedings of the Third Japan – USSR Symposium on Probability Theory. Edited by G. Maruyama and J. V. Prokhorov. VI, 722 pages. 1976.

Vol. 551: Algebraic K-Theory, Evanston 1976. Proceedings. Edited by M. R. Stein. XI, 409 pages. 1976.

Vol. 552: C. G. Gibson, K. Wirthmüller, A. A. du Plessis and E. J. N. Looijenga. Topological Stability of Smooth Mappings. V, 155 pages. 1976.

Vol. 553: M. Petrich, Categories of Algebraic Systems. Vector and Projective Spaces, Semigroups, Rings and Lattices. VIII, 217 pages. 1976.

Vol. 554: J. D. H. Smith, Mal'cev Varieties. VIII, 158 pages. 1976.

Vol. 555: M. Ishida, The Genus Fields of Algebraic Number Fields. VII, 116 pages. 1976.

Vol. 556: Approximation Theory. Bonn 1976. Proceedings. Edited by R. Schaback and K. Scherer. VII, 466 pages. 1976.

Vol. 557: W. Iberkleid and T. Petrie, Smooth S^1 Manifolds. III, 163 pages. 1976.

Vol. 558: B. Weisfeiler, On Construction and Identification of Graphs. XIV, 237 pages. 1976.

Vol. 559: J.-P. Caubet, Le Mouvement Brownien Relativiste. IX, 212 pages. 1976.

Vol. 560: Combinatorial Mathematics, IV, Proceedings 1975. Edited by L. R. A. Casse and W. D. Wallis. VII, 249 pages. 1976.

Vol. 561: Function Theoretic Methods for Partial Differential Equations. Darmstadt 1976. Proceedings. Edited by V. E. Meister, N. Weck and W. L. Wendland. XVIII, 520 pages. 1976.

Vol. 562: R. W. Goodman, Nilpotent Lie Groups: Structure and Applications to Analysis. X, 210 pages. 1976.

Vol. 563: Séminaire de Théorie du Potentiel. Paris, No. 2. Proceedings 1975–1976. Edited by F. Hirsch and G. Mokobodzki. VI, 292 pages. 1976.

Vol. 564: Ordinary and Partial Differential Equations, Dundee 1976. Proceedings. Edited by W. N. Everitt and B. D. Sleeman. XVIII, 551 pages. 1976.

Vol. 565: Turbulence and Navier Stokes Equations. Proceedings 1975. Edited by R. Temam. IX, 194 pages. 1976.

Vol. 566: Empirical Distributions and Processes. Oberwolfach 1976. Proceedings. Edited by P. Gaenssler and P. Révész. VII, 146 pages. 1976.

Vol. 567: Séminaire Bourbaki vol. 1975/76. Exposés 471–488. IV, 303 pages. 1977.

Vol. 568: R. E. Gaines and J. L. Mawhin, Coincidence Degree, and Nonlinear Differential Equations. V, 262 pages. 1977.

Vol. 569: Cohomologie Etale SGA $4\frac{1}{2}$. Séminaire de Géométrie Algébrique du Bois-Marie. Edité par P. Deligne. V, 312 pages. 1977.

Vol. 570: Differential Geometrical Methods in Mathematical Physics, Bonn 1975. Proceedings. Edited by K. Bleuler and A. Reetz. VIII, 576 pages. 1977.

Vol. 571: Constructive Theory of Functions of Several Variables, Oberwolfach 1976. Proceedings. Edited by W. Schempp and K. Zeller. VI, 290 pages. 1977

Vol. 572: Sparse Matrix Techniques, Copenhagen 1976. Edited by V. A. Barker. V, 184 pages. 1977.

Vol. 573: Group Theory, Canberra 1975. Proceedings. Edited by R. A. Bryce, J. Cossey and M. F. Newman. VII, 146 pages. 1977.

Vol. 574: J. Moldestad, Computations in Higher Types. IV, 203 pages. 1977.

Vol. 575: K-Theory and Operator Algebras, Athens, Georgia 1975. Edited by B. B. Morrel and I. M. Singer. VI, 191 pages. 1977.

Vol. 576: V. S. Varadarajan, Harmonic Analysis on Real Reductive Groups. VI, 521 pages. 1977.

Vol. 577: J. P. May, E_∞ Ring Spaces and E_∞ Ring Spectra. IV, 268 pages. 1977.

Vol. 578: Séminaire Pierre Lelong (Analyse) Année 1975/76. Edité par P. Lelong. VI, 327 pages. 1977.

Vol. 579: Combinatoire et Représentation du Groupe Symétrique, Strasbourg 1976. Proceedings 1976. Edité par D. Foata. IV, 339 pages. 1977.